Technology Ethics

Technology Ethics

Responsible Innovation and Design Strategies

Steven Umbrello

polity

First published in 2024 by Polity Press

Polity Press
65 Bridge Street
Cambridge CB2 1UR, UK

Polity Press
111 River Street
Hoboken, NJ 07030, USA

ISBN-13: 978-1-5095-6404-0 (hardback)
ISBN-13: 978-1-5095-6405-7(paperback)

A catalogue record for this book is available from the British Library.

Library of Congress Control Number: 2023923622

Typeset in 11 on 13pt Sabon
by Fakenham Prepress Solutions, Fakenham, Norfolk NR21 8NL
Printed and bound in Great Britain by TJ Books Ltd, Padstow, Cornwall

For further information on Polity, visit our website:
politybooks.com

Contents

Acknowledgments

For almost a decade, I have been interested in and working on the philosophy and ethics of technology. This book not only draws on that experience but it aims to demonstrate how most of the problems that we tend to fixate on concerning our technologies can be reframed if we look at them from the perspective of design. I cannot do justice to all of the colleagues with whom I have had numerous conversations, which have culminated in the work you see here. In many ways, this work is as much theirs as it is mine. Still, it merits mentioning those who provided me with the initial inspiration to work on this topic. Firstly, this project simply could not have been possible without the generous financial support of the Institute for Ethics and Emerging Technologies, which, for nigh a decade, has been increasingly supportive of my work. Likewise, I want to extend my thanks to my dear colleagues at the University of Turin, in particular Graziano Lingua, Antonio Lucci, and Luca Lo Sapio, who have been sources not only of support but also of dear friendship. Mary Savigar and all the production staff at Polity have been invaluable in bringing this volume to fruition. Finally, I would also like sincerely to thank Nathan G. Wood for his impeccable editing skills, which have led to this polished volume.

1

Technology and Society

In 1980 Langdon Winner published what would become a foundational work in the burgeoning field of philosophy of technology. In his paper, "Do Artifacts Have Politics?", Winner described how the parkways of Long Island, New York were built intentionally low (Winner 1980). The reason for this was that Robert Moses, the American urban planner responsible for planning much of New York's metropolitan area throughout the early and mid twentieth century, purposefully designed the parkways low to ensure that poor and lower-middle-class families (mostly African Americans and other minority groups) could not access Jones Beach, one of his prized strands. Moses knew that these groups had limited access to cars and relied on public transit, and those low-hanging parkways could not accommodate tall city buses. The parkways thus created an infrastructural barrier limiting access to Long Island's beaches to only those who could afford cars (Caro 1975). Moses' racist values were thereby embodied in the technology, low-tech as it may be, of the parkways, and this is, in fact, exactly what Winner showed, that technologies are not merely tools, but that they embody values.

Since Winner's work, philosophy of technology has come a long way, and it is now standard to view technologies not as isolated artifacts, but as infrastructures, systems, or, more specifically, as sociotechnical systems. But what exactly does that mean? What does it mean to understand technology as somehow being "sociotechnical"? In both academic and everyday circles, people generally talk about technology in (at least) one of three ways. The first is to conceive of technology purely as a tool or instrument. Usually referred to as instrumentalism, such views are often pushed by those who wish to tout the benefits of a given technology while downplaying possible negatives. A notable exemplar is the oft-quoted motto of American gun rights activists; "guns don't kill people; people kill people." The second way to construe technology is as being purely deterministic. This position, <ism, holds that both human action and our social world are determined by technology, a view nicely illustrated in the popular cyberpunk video game *Deus Ex: Mankind Divided*, where the hashtag #CantKillProgress is repeatedly used to show there is no way to stop the inevitable march of technology and its societal consequences (*Deus Ex* 2011). The third way of looking at technology is to understand it as socially constructed. This position, known as social constructivism, sees technology as being nothing other than the product of human actions; humans, therefore, are completely responsible for what technologies are and what they do. Each of these narratives sees continual propagation in both popular culture and academia, but do they accurately capture what technologies really are?

Robert Moses' bridges show that technologies can both instantiate values and be shaped by them. Moreover, technological limitations can impact how values are embodied in technologies and may alter the very values themselves; interaction effects may stack, interfere with one another, or shift the course of design. All in all, it seems plain that technology is not as simple as any of the single above conceptions would have us believe. Rather, sociotechnicity is a rich yet complex topic in constant development, referring to the dynamic interaction between technologies and people, which form a complex infrastructure (Ruth and Goessling-Reisemann 2019). This means that technologies are not isolated objects. Instead, they are connected systems, part of a larger network of other technologies and people. This sociotechnical understanding of technology highlights a combination of instrumentalism, and social constructivism, and represents what some scholars call interactionalism. Fundamental to interactionalism is the understanding that technologies are in constant and dynamic interaction with other technologies and people.

It may go without saying, but it is also worthwhile to make clear, that technologies provide us with a host of benefits, and we should not automatically assume that all technologies embody disvalues like Moses' racism in his bridges. That example is used to demonstrate that technologies are characterized by the values that they embody and that those values have material impacts on the world and our future alongside them. However, as the world changes, those impacts may change as well; as cars became more affordable, those groups Moses hoped to keep out became more and more able to pass under his parkways and access Long Island's beaches. How a technology embodies a value, therefore, changes over time. This further illustrates how technologies are *interactional*,[1] part of a larger environment of relationships with people and other technologies. Each technology is sure to be designed for an explicit purpose, but they will also interact with other technologies, forming a network of shifting relationships which is important to fully understand if we are to ensure that we design our technologies for good.

Focusing on the values behind development can also be crucial for identifying when a design is failing to fully live up to those values. As an example, artificial intelligence (AI) technologies can illustrate in distressing clarity what can happen when core human values are not clearly and explicitly designed for (Coeckelbergh 2020). For example, IBM spent $62 million USD to develop their famed Watson AI system to help provide medical doctors with cancer treatment advice (Ross and Swetlitz 2018). However, when tested in real-world settings, the system often recommended "unsafe and incorrect" cancer treatments, such as recommending medications that would aggravate, rather than help patients with serious bleeding. Because the data used to train the system was mostly hypothetical, rather than real, the system made poor recommendations. Documents revealed that the decision to use hypothetical clinical scenarios rather than the statistical data of real patients and cases was the consequence of training the system according to the preferences of doctors rather than the Big Data that is available in healthcare, presumably in order for the designers to quickly implement the system. Accuracy and safety were obviously *not* the values explicitly designed for in this system, leading to potentially lethal consequences. There are, moreover, numerous examples where systems have, as a function of design, not only made errors but reinforced existing problems. This is what happens when technologies are not approached from an applied ethics perspective, when we do not look at them as interactional, paying heed to how their various facets impact on one another. Good intentions are not enough; good design is better.

This Book

Technologies, arguably, are an inextricable part of what characterizes human beings, and they are certainly here to stay. Likewise, we are currently experiencing an almost dizzying boom in information and communication (ICT) technologies and artificial intelligence systems that are increasingly difficult to understand (Ihde and Malafouris 2018). If technologies embody the values of their creators, whether they intend to embody them or not, that means that we exert a degree of control over how those technologies impact on our world and the future. This is a hopeful prospect.

This book explores the nuances of how our different sociotechnical systems, systems we often overlook and take for granted, influence and are influenced by our actions. It aims to give the reader a clear overview of how technological design has been traditionally handled, how and why philosophy has become so important in design, as well as the various approaches for actually doing the dirty work now so that we don't suffer the consequences later. More broadly, this book will introduce philosophical concepts and positions as they relate to how we understand technologies and our relationship with them, while also showing how important it is for engineering ethics that we have an accurate and holistic understanding of technology.

Towards this end, this book will explore some of the main historical and current views of technology, as well as connect philosophical concepts to practical applications. This will help guide readers in understanding the importance of engineering ethics, that is, understanding and promoting the ethical practice of engineers (Harris et al. 2013). Given the ubiquity of technologies in our hyperconnected world, and given the role that engineers play in the creation of those technologies, understanding and promoting engineering ethics is an important goal. Doing so requires people from various disciplines and fields like philosophy, public policy, and, of course, engineering, to come together. Huge investments at regional levels, like that of the European

Union, moreover demonstrate the overall interest in promoting this practice.

Focusing on what technology is and what engineers can do to ensure that technologies are designed and developed ethically means that we can focus more on pressing real-world issues that come as part and parcel of technologies and less on the techno-utopian or techno-dystopian narratives that have been dominant in both public as well as academic spaces. Many scholars that have directed their energy toward engineering ethics have found that those latter hyperbolic debates often come at the opportunity cost of more proximal issues that contemporary technologies present and that need immediate attention, like unemployment as a consequence of automation or issues of data privacy. That, of course, does not mean that thinking about the long-term future should be totally sidelined, as doing so would risk missing the forest for the trees, thinking too narrowly when designing and introducing new technologies into the world. For this reason, in the last chapter, I discuss engineering ethics as a multi-generational project, thinking about design as something that necessarily transcends individual lifespans.

Likewise, it is valuable, philosophically speaking, to consider narratives that are beyond those of exclusively extant technologies. Science fiction and fantasy provide us with modalities of understanding the complexities of the world in novel and complex ways. In his book, *New Romantic Cyborgs*, Mark Coeckelbergh explores how science fiction narratives concerning robots impact and influence how we actually perceive such technologies (Coeckelbergh 2017). In this book, I regularly draw on fictional analogies and narratives, such as from the works of J.R.R. Tolkien, to help tease out what would otherwise be complex and nuanced philosophical positions as they relate to how we can properly understand what technologies are.

For example, the Ring of Power in Tolkien's lore is a testament to the creator's will, embodying the immense influence it holds over both its lesser counterparts and their bearers.

> Now the Elves made many rings; but secretly Sauron made One Ring to rule all the others, and their power was bound up with it, to be subject wholly to it and to last only so long as it too should last. And much of the strength and will of Sauron passed into that One Ring; for the power of the Elven-rings was very great, and that which should govern them must be a thing of surpassing potency; and Sauron forged it in the Mountain of Fire in the Land of Shadow. And while he wore the One Ring he could perceive all the things that were done by means of the lesser rings, and he could see and govern the very thoughts of those that wore them. (Tolkien 2007, 265)

Such a narrative device reflects the potential of technology to not only command over other creations but to embed itself into the very fabric of societal functioning, influencing actions and decisions. For example, the 1999 film, *The Matrix*, can be a vehicle to help us think about how modern media technologies can determine and cause cultural change.

For the purposes of this book, these narratives can help us follow how different characters view technologies and how that impacts on their choices and actions. How do the different characters in Tolkien's magnum opus, *The Lord of the Rings*, view the *One Ring*? Does it determine their behavior? Can it be controlled and used towards good ends? As modern scholars of technology have argued, the various ways of conceptualizing the same technology can lead to radically different outcomes. For engineers and designers who have yet to take applied ethics seriously in their day-to-day work, drawing on a narrative like that of *The Lord of the Rings* may provide a more nuanced point of departure for navigating away from simplistic and facile ways of understanding what technology is and towards a more complex and nuanced understanding that can address many of the difficult and seemingly intractable issues we face today.

What happens when we take a more nuanced understanding of technology that moves beyond the simplistic narratives of instrumentalism, determinism, and constructivism?[2] Viewing technology as something that is inextricably co-constitutive of everyday human life, but one which can be guided and directed, means that many of the ethical issues resulting from technological development can (potentially) be addressed by conscious design. This means that not only can engineering ethics be informed by various more foundational ethical perspectives and concerns, but that engineering ethics can likewise inform ethics itself, with each providing impetus for development in the other. What we'll see throughout this book is that each of the traditional ways of understanding technology misses something. Instrumentalism misses values; determinism misses humans; and constructivism misses how technologies impact on our lives. Each of the philosophical positions presented in this book has its merits, but also its shortfalls. One of the threads that weave this book together is that technology is best understood as embodying values; tools influence our behaviors, and their design is influenced by our actions. Taking this as a point of departure, engineers can better understand how their decisions impact on ethical issues and reflect on how to mitigate any unwanted consequences. Philosophers can look at how technologies bring to light novel and unforeseen ethical values. Together, they may ensure that emerging technologies bolster our most fundamental and sacred human values.

It is also worth stating that the treatment given to the philosophical positions and approaches in this book is limited and has been described in greater detail in other works. For example, much more could be said regarding the design of artificial intelligence systems, and the issue of engineering education, that is, teaching engineering ethics, is not given any attention. The primary aim of this book is to serve as a foundational text for students and professionals in the fields of philosophy of technology, technology ethics, and Science, Technology, and Society (STS) studies. It is designed to introduce a wide array of readers – including philosophers, designers, and policy-makers – to the multifaceted ways of engaging with technology and the ethical considerations that arise in conjunction with their design and application. I have thus decided to limit the purview of this book to how we can properly conceptualize technology, why that is important, and the approaches that engineers can take in order to guide technological development towards beneficial ends. (The interested reader is directed to more intensive treatments of these subjects in the Further Reading section of this volume.) In building toward these points, we will begin with the existing – and flawed – conceptions of technology, and work our way toward a more fruitful and philosophically well-founded conception. With a clearer picture of these foundational points in mind, we will then consider the more hands-on task of actually doing technology ethics. The book is thus structured as follows:

Chapter 2 examines one of the most pervasive and historically rooted conceptions of what technologies are; i.e., instrumental. What does it mean to view technology as purely instrumental? Can you treat technology, like a gun, as nothing other than a tool that can be used for good or evil? Are technologies simply part of a continuum of development and improvement? What are the ethical and practical implications of viewing technologies as instrumental, as solutions to other technological problems?

Chapter 3 explores a narrative more common in popular media, namely technological determinism. Here we explore the idea that technology is best understood as the determinant of our behavior. This means that technologies guide human history and that human actions cannot impact on technological development. What are the different ways of viewing technological determinism? Are all technologies deterministic, both past and present? Who promotes such narratives, and should technologies guide our behavior? What are the limits of technological determinism?

Chapter 4 looks at the foil of technological determinism, namely social constructivism. This position grew out of a criticism of technological determinism and argues that technology is best understood as a by-product of human behavior and decisions, and that there is no internal logic inherent to technology that guides our behavior. What are the different ways in which technologies are socially constructed? Are there benefits of viewing technology as socially constructed? What are the critiques of social constructivism of technology, and what can we learn from them?

Chapter 5 brings together the advantages and drawbacks of the previously discussed views, building toward the so-called design turn in applied ethics, and arguing that technology is best understood as interactional, a mix of instrumentalism, determinism, and social constructivism. This approach to understanding technology is aspirational and looks at ways to promote well-being through design. But it leaves open important questions as well. What does it mean for technologies to "embody" values? What does it mean for technology to be "interactional," and why should we view it like that? What does it mean for designers and engineers that technology is interactional?

Chapter 6 looks at responsible innovation as a concept, the broader movement, and the concept that emphasizes the responsibility of the people involved in creating innovations. A large amount of funding and attention toward responsible innovation betrays the need to guide the potential impacts that technologies ultimately have when they are introduced and diffused in society. But what is responsible innovation? How do we go about responsibly innovating? And what are examples of responsible innovation, and how successful are they?

Chapter 7 takes a closer look at a few ways responsible innovation can be pursued. Various approaches to responsible innovation are proposed, like universal design, inclusive design, participatory design, and value sensitive design. In examining these, we ask what benefits are brought by the different approaches to responsible innovation. More specifically, are the different approaches good for every application? Are there any limits or shortfalls of these approaches? Are they tried and tested?

Chapter 8 goes into the nuts and bolts of responsibly innovating, presenting an "ethicist's toolbox" that designers and engineers can use in their day-to-day work. In this chapter, we look at specific cases of responsible innovation in practice for each of the approaches discussed in previous chapters. This gives practitioners a quick look at how to put abstract theory into practice. Philosophers may also gain insight here by seeing how ethics becomes operationalized and put into use in design approaches.

In the final chapter, Chapter 9, we explore the concept of "progress, rather than perfection" as the driving force for technology ethics, highlighting the idea that ethics in technology is a process of continual examination; there is no perfect endpoint. More than this, searching for perfection can present dangers of its own and stifle necessary beneficial progress. How should we approach technology development in a continually changing society? How do we confront technologies that learn and change over time? In this last chapter, the book aims to remind the reader that humans and their technological creations are constantly in a process of co-creation, meaning that responsibility is a lifetime duty; we cannot simply make technologies, throw them into the world, and wash our hands of them. Technologies are too central and too dynamic for us to simply abdicate our responsibility for what they do in the world.

2

Instrumentalism

Technologies as Tools

There is a sense in which it is natural to think of technologies as mere tools which can be used at the discretion of their bearers. On this view, technologies are essentially value neutral. Thus, when thinking of a simple technology like a hammer, in the hands of one individual it may be used to drive nails into wood, but another may use it as a lethal weapon. However, there is nothing about the hammer itself that necessarily makes it a tool for carpentry or a murder weapon. This instrumental understanding of technology is one that is almost ubiquitous in everyday thought, and it is far from new. In fact, the idea of technologies as tools goes back to (at least) the ancient Greek myth of Prometheus, where the "tool" of fire is given to humanity, enabling us to create additional tools and reach ever higher levels of civilization (Raggio 1958), and it is hard to even separate an understanding of technologies as tools from what it means to be humans.

Following in the tradition of the story of Prometheus, humans are more than simply *Homo sapiens*, but *Homo faber*: "Man the Maker." Through the use of technologies (as tools) humans have made themselves capable of controlling their evolution and environment

(D'Amore and Sbaragli 2019), and even today, this is a common understanding echoed by many self-described transhumanists, who argue that technology can and should be used as a means to further human evolution (Bostrom 2005). It is no wonder that such a position, longstanding and pervasive as it is, is so common in both popular and intellectual thought, and for good reason. However, the merits and intuitiveness of viewing technologies as tools has limits, and sooner or later, it becomes necessary to view technologies more broadly.

There is another sense in which it seems natural to construe life as part and parcel of technology development, which is true to a large extent; technologies have indeed inextricably mediated how we live our lives and communicate with one another. The technologies that have been developed over the course of human history can undoubtedly be viewed as a continuum. The bow and arrow is certainly the successor of the atlatl, as the first firearms succeeded the bow. One might argue that the development of technologies is primarily one of degree rather than kind. However, things are not as simple as this. (Chapter 5 will explore how technologies are also best understood as forming part of a more extensive infrastructure.) Still, it merits noting both the boons and pitfalls of viewing technologies as tools as well as how it continues to be a common narrative approach to viewing technologies more broadly.

Guns

One of the primary aspects of instrumentalism is the understanding of technology as a neutral tool, a position often referred to as the neutrality thesis (Peterson and Spahn 2011). The neutrality thesis holds that tools have no inherent (dis)values, and are only to be seen in instrumental terms. Thus, while it may be perfectly acceptable to use a hammer for driving nails or utterly reprehensible to use a hammer to kill a man, there is nothing in the hammer that makes it good or bad. A more prominent example of the neutrality thesis is the oft-pushed position of the National Rifle Association (NRA) that "guns don't kill people; people kill people" (Henigan 2016). The NRA, an American lobby group, supports and promotes individuals' rights to bear arms, and they bolster their position with the idea that technologies (guns) do not embody values, but instead are neutral artifacts which can be used in different ways according to the intentions of their users. In practice, this means that whether or not a gun is a "good" thing depends on how it is used, which in turn depends on the user's intention(s). If the user has malicious intentions, the gun is a tool for bad ends, but by the same token, a virtuous person can use a gun for virtuous ends. Whether it is used for good or bad, the gun remains the same. In fact, this exact line of reasoning is used when confronting the public backlash following gun-related tragedies, as gun rights advocates are quick to point out that the "only thing stopping a bad guy with a gun is a good guy with a gun" (BBC 2012). The point is cleanly captured by Fabio Tollon, who argues that

> The suggestion here (from the NRA at least) is that if we can learn to simply be better persons, then we do not have to worry about the moral effects of artifacts. If we are trained, for example, to uphold better gun safety standards, etc. then we would have done all we can. [...] nobody would claim that the gun makes no contribution to the killing, and nobody would claim that the gun is wholly responsible either. Those who oppose the proliferation of guns merely assert that these artifacts can affect those who make use of them. Conversely, gun control opponents merely claim that guns are but one efficient way of carrying out an act, with other things also capable of performing the same task. (Tollon 2022, 240)[3]

In Japan, the idea of technology as a neutral tool is less prevalent when it comes to firearms. The Japanese society and legal system have a different view on the neutrality thesis, especially regarding guns. The Japanese *Swords and Firearms Possession Control Law* is extremely strict, and it is based on the premise that the possession of guns inherently increases the risk of harm (Alleman 2000). This perspective challenges the neutrality thesis by suggesting that the mere presence of a tool, in this case, a gun, can influence behavior and societal outcomes. The law in Japan is not just about controlling the use of guns but also about limiting their availability. It is a reflection of a societal belief that guns are not merely neutral tools that can be used for good or bad depending on the user's intentions. Instead, it suggests that the very existence of guns in society can lead to negative outcomes, regardless of individual intentions. This perspective is supported by Japan's low gun-related crime rates, which are among the lowest in the world (Karp 2018).

Now, to a certain extent, guns, like all technologies, do have an instrumental aspect to them. There is a sense in which they are "merely tools," and in fact there are numerous instances of guns being used to stop "bad guys," just as there are numerous instances of "bad guys" using guns for murderous ends. The gun remains a gun. Yet as we shall see in later chapters, characterizing technologies in this way, as *purely* neutral, mistakenly overlooks a more foundational and nuanced way of understanding technologies, one that is partly instrumental but also charged with values.

Only Kinda Neutral

There are certainly limitations to viewing technologies as nothing more than neutral tools devoid of values.

Thinking back to the previous chapter, the classic example of Long Island's parkways demonstrates just how value-laden technologies can be, and the recurring example of Tolkien's *One Ring*,[4] which we will see throughout this book, paints a stark illustration of the potential interconnectedness of technologies and values. Fundamentally, we simply must accept that there are multiple facets to technology, and a more holistic understanding must draw on various aspects of instrumentalism, determinism, and social constructivism (the latter two will be discussed in greater detail in Chapters 3 and 4).

In addition to the lack of appreciation of values, instrumentalism also ignores the ways technologies can impact on one another, how they may permit, mediate, or constrain certain other technologies. For example, nails beget hammers, tools that necessarily pair together and continue to be used today. The introduction of electrical and motor technologies allowed the hammer-nail system to be further enhanced, building upon the rudimentary technologies but to a greater degree. Likewise, computer-controlled production systems employing automated nailing machines push this even further. In short, the simple technologies of hammers and nails set a foundation upon which more sophisticated systems could be built, up to and including automated devices which are more convenient, less laborious, and more efficient than their early predecessors. Naturally, the evolution is one of degree rather than kind, but it betrays a seamless connectivity between numerous technologies, and shows that technologies must be understood in more holistic and infrastructural ways. When we mention one technology, it becomes impossible not to mention the others that are inextricably connected to it. More than this, technologies inextricably link to, impact on, and are affected by human behavior, a thought cleanly captured by Nicola Henry and Anastasia Powell:

> Thus, technologies are not understood as neutral (a mere addition to a pre-given social system), or determinative (directly causal of changes in a social system) but as an embedded and co-constituting feature of society and its structures, cultures and practices. (Henry and Powell 2016, 36–37)

This means that our tools are not simply bearers of our intentions but rather mediators and partial determinants of what those intentions can be. A hammer can drive a nail, just as it can kill. However, the same hammer eliminates other possibilities, like that of being a tool for eating soup, writing a novel, or calculating the square root of a large number (Verbeek 2005, 114). This, we shall see later, is why design is essential. Although designers can never foresee all of the possible ways their technologies can be used, they can anticipate (to a degree) how their technologies might impact on society, how society could affect the use and understanding of those technologies, and how those technologies may open or close future technological possibilities.

Down but Not Out

Viewing technologies as purely instrumental has largely fallen out of favor in academic discussion, but it has not disappeared entirely. This is a problem, as the instrumental view can, albeit unintentionally, serve to manufacture consent for inequitable power structures, due to the fact that it ignores technologies' systemic and connected nature. For example, *We Have Always Been Cyborgs*, by Stefan Sorgner, takes a close look at biomedical technologies such as in vitro fertilization and preimplantation genetic diagnosis, and in doing so considers these technologies in much the same way that the NRA views guns; the technology of one domain is seen as being potentially applicable to another domain, without any impact coming from the original intent and purposes of the designers (Sorgner 2021, 15). So something like predictive health maintenance may come with boons, as Sorgner suggests, but it can just as easily be used as a precondition for invading privacy or gaining control over health care, a fact which has already seen precedents (Wakefield 2022; Stamouli 2022). More generally, technologies intended for one purpose can be appropriated and used for others, but even more starkly, the design of technologies may support and/or constrain the ways technologies can be used and the domains to which they may be applied.

What Sorgner's arguments and positions betray is that he is using an instrumentalist understanding of technology. This way of conceptualizing technology, and its multi-use nature, leads Sorgner to encourage the "techno-fix" approach to problem-solving. More specifically, he argues that many of the global challenges that we face, like climate change, can be addressed with other high technologies like in vitro meat, solar panels, novel architectures, and novel modes of transportation (Sorgner 2021, 15). Instrumentalism often leads to a unidimensional understanding of problems by providing low-resolution answers. Many of the causes of increasing global temperatures, at least from the anthropic side, are the direct and mostly unintended results of high technologies.[5] There are also often hidden costs when using techno-fix solutions, and these costs require their own solutions in turn.[6] Thus, the proposed solution can lead to more problems of the exact kind meant to be addressed (Umbrello 2023b).

In their book, *Techno-Fix: Why Technology Won't Save Us or the Environment*, Michael and Joyce Huesemann explore the marvels that modern technologies have ushered in on a global scale (Huesemann and Huesemann 2011). However, the conveniences and benefits brought by these technologies come with many unforeseen and perhaps unforeseeable consequences as well, like anthropogenic climate change. Current research into climate change (and its amelioration) has led to a range of potential solutions, both on the policy and technological sides, and the latter, geared as technofixes, generally come under the umbrella of climate engineering, a technology family that includes techniques for carbon dioxide removal and solar radiation management (Buchinger et al. 2022). Yet these "solutions" can (and arguably will) give rise to further problems.

Like the technologies that caused the issues in the first place, the unforeseen (and possibly unforeseeable) side effects of technologies often do not appear until those technologies are firmly entrenched in society, and at that point, it can be difficult to rein in the use and impacts of those technologies. This concern is often called the *Collingridge Dilemma*, where we cannot know the effects of a technology until it has been implemented in society, but once it has been implemented, it becomes difficult (if not impossible) to later remove that technology from society (Genus and Stirling 2018). Using global technofixes like climate engineering technologies also carries the danger that unintended side effects will likewise be global in scale (Morton 2015). The complexity of Earth's climate system makes it hard, if not impossible, to predict all of the implications of a particular climate intervention, meaning that potential solutions like stratospheric aerosol interventions, where gas is injected into the atmosphere to convert into solar radiation-blocking aerosols, may ultimately cause more harm than good, substituting one intractable problem for another (Adomaitis et al. 2022).

This way of thinking pervades naturally even into policymaking, where policies take instrumentalism and value-neutrality as a given when drafting recommendations. "Net Neutrality Policies," for example, are based on the idea that all internet traffic should be treated equally, without any discrimination or preference given to certain types of content, sites, platforms, applications, or users. This assumes that the internet infrastructure is a value-neutral technology that merely facilitates data exchange without influencing the content or shaping user behavior. China, however, takes a wholly different approach here. In China, the government's approach to internet policy is far from the idea of net neutrality. The Chinese government has implemented a system known as the "Great Firewall," which actively discriminates against certain types of content and platforms. This system blocks access to foreign websites that are deemed inappropriate or harmful by the government, including popular Western platforms like Google, Facebook, and Twitter (Sari et al. 2017). Instead, Chinese users are directed towards domestic platforms that the government heavily regulates.

The Departure of Boromir

The optimistic view of technology as the potential solution to global problems ignores the fact that those problems are part and parcel of developing and using other technologies. Moreover, this uncritical acceptance of technofixes as the path to solving our problems – a view which is still characteristic in many academic and public discourses – is underpinned by the value-neutral, instrumental, conception of technology. Yet this conception overlooks much, to its detriment.

Returning to the running example of this book, the *Fellowship of the Ring* makes the naivety of the instrumental view quite clear. Boromir, seeing the ring as nothing more than a tool, ardently suggests harnessing its might, saying, "Yet may I not even speak of it? For you seem ever to think only of its power in the hands of the Enemy: of its evil uses not of its good. The world is changing, you say. Minas Tirith will fall, if the Ring lasts. But why? Certainly, if the Ring were with the Enemy. But why, if it were with us?" (Tolkien 2004, 398). The *One Ring* is an artifact of great power, but it is a power that was made *for a purpose*. Seeing the ring as nothing more than a tool, Boromir argues that

> And behold! in our need chance brings to light the Ring of Power. It is a gift, I say; a gift to the foes of Mordor. It is mad not to use it, to use the power of the Enemy against him. (Tolkien 2004, 398)

Boromir is right that the ring has power, and the temptation to use that power, even for noble ends, has intuitive merit. Yet Boromir falls into the same trap as all those who try to disentangle a technology from its purpose. The *Ring* was made as a tool of domination, and though it likely could be used instrumentally to depose the enemy, in so doing, it would utterly corrupt the user who claimed it as their own. The values of the ring are bound up within it, and to ignore those values is to ignore inevitable realities which are tied to that device.

Going forward, we will see how this analogy provides an illustration of technology more broadly as being incarnate of values. At any rate, and as we shall see in the following two chapters, viewing technology as purely instrumental often sidelines important facts, and though the instrumental *side* is a side of technology, it cannot paint a complete picture. The next chapter will examine technology's deterministic impact on society and our behavior. However, it will also highlight that viewing technology as exclusively deterministic brings with it its own set of problems.

3

Technological Determinism

> Those who do not put in the time and effort to understand Technological Determinism are determined to repeat it.
>
> *David Gunkel*[7]

The Inevitable March of Progress

In the popular video game *Deus Ex: Human Revolution*, players are thrown into a techno-dystopian world where humanity is separated into two distinct groups, those who possess artificial organs (the augmented) and those who do not. A technological apartheid results. The advances in biotechnology and cybernetics that form the game's backdrop are described as being part of the inevitable power of progress, an unfolding of history that can be neither guided nor stopped. The social divide between the "augs," those with the desire and necessary wealth to obtain and sustain augmentations, and those who do not, is framed as a natural condition of technological progress. The central notion that technological progress (and its attendant impacts on society) is inevitable forms the foundation of a class of thought called *technological determinism*, a position captured quite nicely in the marketing campaign of the game, where across social media platforms, it was promoted with the hashtag #CantKillProgress (*Deus Ex Wiki* n.d.).

Technological determinism, simply put, is the idea that technology directs and changes fundamental aspects of our behavior and our society; technologies direct how we relate and interact with one another, what our values are, how we understand those values, and even shape how we learn, think, and conduct ourselves in broader society (Kline 2001). We can see this type of narrative in the UK's National Data Strategy, which describes data as "the driving force of the world's modern economies," treating data itself as the driver of change and intent (Department for Digital, Culture, Media and Sport 2020). Neil Postman, a media theorist and cultural critic, argued from an explicitly determinist position that accepting technologies in society amounts to what is effectively a "Faustian Bargain" (Postman 1995, 2:08 min), as each technology

> gives us something important, but it also takes away what's important. That's been true of the alphabet, the printing press, and telegraphy right up through to the computer. When I hear people talk about the information superhighway; it will become possible to shop at home and bank at home and get your texts at home and get your entertainment at home, and so on, I often wonder if this doesn't signify the end of any meaningful community life […] that human sense of responsibility we have for each other. (Postman 1995, 2:08–3:26 min)

Implicit in this position is a sort of value neutrality, an issue already discussed in the previous chapter in relation to instrumentalism. What we see in Neil Postman's position, something that is often found in technological determinism, is that the value of the technology *per se* is not even discussed, as the focus is instead entirely on how technology impacts on society

(Harrison n.d.). Andrew Feenberg fiercely sums up the technological determinists' understanding of technology as a "two-sided phenomenon":

> on the one hand there is the operator, on the other the object. Where both operator and object are human beings, technical action is an exercise of power. Where, further, society is organized around technology, technological power is the principle form of power in the society. This is its dystopian potential. (Feenberg 2001, 141)

Both sides of this phenomenon construe technology as the driving force in society, as power. But this alone is a rather broad conception of technological determinism, and it merits looking at how the internal logic of specific varieties of the theory work before we move to critique them.

Flavors of Technological Determinism

There are a number of different contemporary ways that technological determinism can be viewed, but all have their origin in a nineteenth-century blending of an industry/crafts conception of technology and an anthropological understanding of technology as "artifacts and related practices of a culture" (Kline 2001, 15495). But this early paradigm quickly evolved into the starker notion of technological determinism that we still see today.

Given the rise of automation and the ongoing debates surrounding Karl Marx's theory of history as a form of "economic determinism," scholars began searching in the twentieth century for a way to guide the immense and seemingly unstoppable impact of various new technologies on society (Marx 1994; Heilbroner 1967).

However, for all of the guidance that might be achieved, one fact that remained was that technology was the force pushing society (and those scholars seeking to shape technology's impact).

In distinguishing between types of technological determinism, Bruce Bimber picks out three different accounts: (1) normative accounts, (2) nomological accounts, and (3) unintended consequence accounts (Bimber 1990). In the normative accounts, Bimber argues that social norms are sidelined and replaced by norms of technology. More precisely, society gives up and replaces political norms with the objectives of efficiency and productivity, goals characteristic of technologists. This normative position can be found in the works of technology critics like Jürgen Habermas and Lewis Mumford. Nomological accounts, on the other hand, treat technological determinism in the same way we would laws of nature; technologies are understood as possessing internal forces which determine the technologies themselves and shape social change. The unintended consequences account takes a different path, arguing that technology produces unforeseen social effects which are (seemingly) partly random and are beyond human control (Winner 1977). Of these three positions, the first and the third are arguably not quite deterministic enough (Bimber 1990), and true technological determinism is best captured by nomological accounts, where it is the internal force of technology that determines social change. Yet, even so, each view captures an aspect of how technology may impact on society.

From the Stirrup to Mass Media

On the nomological version of technological determinism, certain contemporary systems like artificial intelligence most clearly exemplify technology's internal forces, which can drive social change. However, we can see this going all the way back to medieval Europe, where a relatively simple device demonstrated just how significantly technology can shape society.

In *Medieval Technology and Social Change*, Lynn White argues that the introduction of the stirrup, a frame attached to a horse's saddle which holds the rider's foot, was the determining factor for the emergence of feudalism in Europe (White 1962).[8] White argues that stirrups allowed for more effective mounted shock combat, which in turn aided the military class in more efficiently buttressing local feudal structures. This was due to the fact that the mechanics of the stirrup

> made possible – though it did not demand – a vastly more effective mode of attack: now the rider could lay his lance at rest, held between the upper arm and the body, and make at his foe, delivering the blow not with his muscles but with the combined weight of himself and his charging stallion. (White 1962, 2)

Though shock cavalry was by no means a new concept, this technology improved it greatly, and provided what White argues was a "diversion of a considerable part of the Church's vast military riches ... from infantry to cavalry" (White 1962, 4), given the degree of social control that it directly provided via the feudal system. The stirrup, then, according to White, was the driving technology directly responsible for subsequent feudal development across medieval Europe.

Although White's account has not gone uncriticized – it sparked the "Great Stirrup Controversy" (Farndon 2010) – it nonetheless provides a keen analysis of the deterministic power that technology can and does have on our behavior and society. Moreover, though the stirrup may be far from commonplace for most of us in the modern world, the same point can be seen with a contemporary technology which determinists view as the hallmark of the deterministic powers of technology: the media.

Examining the media and its impact on society, Marshall McLuhan explored a particular form of technological determinism now known as media determinism (McLuhan 1994). Like technological determinism, media determinism argues that the media has the power to shape society (De la Cruz Paragas and Lin 2016). Importantly, media determinism moreover posits that it is the *medium of communication* – in this case, the media – which shapes society, rather than the message itself, giving rise to McLuhan's famous saying that "the medium is the message" (McLuhan 1994). Philosophically, McLuhan makes a connection between how language impacts on and shapes our perception and the media's impact on language. As such, media determinism holds that the media's ability to impact language gives rise to its ability to shape and direct how people think and, thus, behave. In a simple example, McLuhan writes that

> [t]he instance of the electric light may prove illuminating in this connection. The electric light is pure information. It is a medium without a message, as it were, unless it is used to spell out some verbal ad or name. [...] a light bulb creates an environment by its mere presence. (McLuhan 1994, 19)

What we see in McLuhan's words is what Martin Hirst describes as a focus on the consequences of a technology itself, and the effects it produces. As a result, in the case of the light bulb, technological determinism invites us to conceptualize "the machine itself [as] appear[ing] to be 'alive,' or at least capable of directing human behaviour" (Hirst 2012, 4). To a certain degree, this is true though; technology, as we shall see, does have a determining power over human behavior. However, how much power any given technology actually has can be debated, and arguably no technologies are properly *determinative* over human behavior or society.

Critiques of Technological Determinism

Though deterministic thinking saturates much popular culture and academic work, there have been notable shifts in thought concerning the limits of determinism. In fact, by the middle of the twentieth century, there was a marked rise in skepticism about deterministic narratives, due primarily to worries about the exculpating effect that technological determinism could have; following World War II and the technologically mediated horrors of atomic weapons and experimentation on human beings by Axis scientists, scholars realized that deterministic narratives seemed to remove responsibility from the people engaging with those technologies. After all, if technology truly was "deterministic," if you #CantKillProgress, then the experimentations of Josef Mengele, the Nazi "Angel of Death," were nothing more than the inevitable outcome of the technologies he used when experimenting on people.

What the objections to determinism highlighted is the question of "should." That is, *should* media be the driving force of human behavior? *Should* increased automation force large swaths of the workforce into unemployment? *Should* our greatest minds be tasked to developing the most terrible of weapons? These are serious questions, as narratives supporting technological determinism also have a tendency to mitigate human responsibility, if not eliminate it altogether. This means that when something goes wrong – which almost always happens at one point or another – those responsible are shielded from being scrutinized for the impacts of the systems they designed and deployed. Sally Wyatt aptly describes this phenomenon, saying that technological determinism

> remains in the justifications of actors who are keen to promote a particular direction of change, it remains as a heuristic for organizing accounts of technological change, and it remains part of a broader public discourse which seeks to render technology opaque and beyond political intervention. (Wyatt 2008, 176)

In other words, technological determinism can be weaponized as a narrative, shaping how technology is developed and perceived, rendering itself a self-fulfilling prophecy. The consequences of this can be catastrophic; early wars were fought with swords, spears, and bows, weapons requiring close degrees of proximity to be effective, but the tide of technology, along with the introduction of gunpowder and firearms, naturally shifted the ways we fought. Likewise, the advent of nuclear power and weaponization of nuclear energy would, according to technological determinism, impact on how future wars will be conducted. Yet this reductivist view treats humans as nothing more than slaves to the ebb and flow of technological change, as incapable of acting on our world to shape the ways technology develops (Winner 1977, 88).

In a related vein, not only can determinism seemingly push humanity toward certain horrific ends, it can also act as a massively constraining force for individual actors in society. For example, Jonathan Benthall argues that the view that technology determines our behavior can, in fact, politically disempower us, leaving us feeling trapped by its power; "today even the most highly placed managers represent themselves as innocent victims of a technology for which they accept no responsibility and which they do not even pretend to understand" (Benthall 1976, 241). The upshot of all of this is that simply conceptualizing technology as deterministic can lead to a self-fulfilling prophecy where people give up their control and responsibility, thinking that they would otherwise have no material impact on the direction of innovation. A good example of this impact is in policymaking where standardized testing is concerned, particularly those that are accompanied by digitalization and data-driven decision-making. The deterministic perspective here assumes that technology – in the form of online platforms, learning analytics, and standardized test formats – can objectively measure student performance and learning outcomes, and therefore improve educational quality and fairness. What this deterministic bent in the policy does, then, is oversimplifies the complex nature of learning, reducing it to quantifiable metrics that may not fully capture a student's abilities or potential. This might lead to a narrow focus on teaching to the test, possibly neglecting other vital aspects of a holistic education, such as critical thinking, creativity, and social skills.

Still others, like Andrew Feenberg, take a different approach to confronting technological determinism, arguing that technology is fundamentally good for people and society. Sometimes referred to as "technology evangelism," Feenberg's position (which mostly refers to the impact of the internet on society) argues that users of the internet (or other technologies) are not only consumers but producers also. As a result, the internet as a medium of communication becomes democratized, an inherent good. This democratization also implies that users can dictate the development of the technology, *pace* determinism, and that this can lead to increasingly good ends.

The Social Construction of Technology

In the next chapter, we will discuss the "social construction of technology," or, more simply, *constructivism*. However, it merits briefly noting here that this view arose mostly in response to the shortfalls of technological determinism and, what Timothy Snyder calls the "politics of inevitability" (Snyder 2018).

Constructivist researchers take a wholly different approach than proponents of technological determinism, arguing that *not* only does technology not shape society, but rather that it is society and human behavior that almost entirely determines (or constructs) how technologies are developed. Essentially, constructivists promote what could be thought of as *social determinism*, highlighting the roles of those various human cogs in the larger machine that leads to innovation; engineers, managers, executives, and users, among others, are, after all, the causal factors determining how a technology actually turns out.

Ultimately though, however one builds an objection to determinism, the main lesson should be that despite those instances where technology truly determines our behavior, we ought nonetheless to be wary of the dangers that lurk in such narratives. There are numerous examples where proponents of the determinist view are using it to their own benefit, concentrating power and influence, all the while washing their hands of their responsibility for innovations they made. This is not to say that technologies do not have a partly deterministic character to them – they rather obviously can and do – but rather to highlight that conceiving of technologies in *purely* deterministic ways leads to serious worries. Moreover, as we will see in Chapter 5, the determinative aspect of technology is not ultimately the result of some internal force or logic of a technology itself, but rather a consequence of its overall design, something which we, as agents, can control.

4

Social Constructivism

Technologies are What We Make Them

In the preceding chapters, we explored two common conceptions of technology. Both brought forward important points worth keeping in mind, but their single-minded adherence to a unary view led both to significant problems as well. The objections raised against *technological determinism*, in particular, have led to the rise of another popular way of conceptualizing technology (which is the focus of this chapter), namely, the *social construction of technology*. In a nutshell, the *social construction of technology* – sometimes called *social determinism*, or simply *constructivism* – claims that it is not technology that influences human behavior, but rather the other way around; humans design technologies for purposes they themselves set, and use those technologies in certain ways determined by their societies and cultures. Constructivists additionally claim that it is impossible to understand how technology is utilized without also understanding how it is incorporated into a given social environment.

Fundamentally, constructivists maintain that in order to know why technology is accepted or rejected, one must examine the social landscape. According to their view, it is insufficient to attribute a technology's success to its status as "the best," akin to some naive notion of Darwinian fitness. Instead, one must first consider how the standard of "the best" is established at all, and who the key players and groups are that contribute to that definition. In particular, researchers must know who establishes the technical standards by which success is judged and why those standards are established. The myth of technological determinism arises only when, looking back, one assumes that the road we've taken to arrive at the present was the *only one* that could have been taken.

Viewing technologies as social constructs further brings up fresh perspectives on how technologies emerge and evolve; it makes it obvious that there are no good or bad technologies; the lines between success and failure, as well as between truth and untruth, are blurred; and the individuals participating in the process of innovation fundamentally determine how a system is shaped. In stark contrast to technological determinism, for the constructivist, technology is not a given. Depending on those doing the innovating, it may take on varied shapes, and a certain technology may appear differently if additional or different actors take part in the design process. In other words, we must look not just at systems themselves, but also at the people behind those systems.

The Principle of Symmetry

In a foundational article for both the sociology of science, and of technology, Pinch and Bijker adopt what they call the *Principle of Symmetry*, which maintains that researchers should use the same types of justification and reasoning when examining both successful and unsuccessful models, ideas, or experiments (Pinch and Bijker 1984). Put differently, researchers should be unbiased about whether a belief is true or not when examining it, and the same goes for explanations. By incorporating this principle, constructivism takes a relativist or neutral stance regarding the justifications offered for the acceptance or rejection of a technology. All arguments – social, cultural, political, economic, or technical – should be given the same consideration.

The principle of symmetry was introduced to address a common issue in the sociologies of both science and technology, namely that historians were prone to explain successes by citing "objective truth" or intrinsic "technical superiority," while casting failures as being the result of political influence or economic factors. However, such fickle choosing of explanations serves to obfuscate and confuse certain realities. Pinch and Bijker, for example, looked to the history of the bicycle to illustrate the point. They argued that the success of the "safety bicycle" – i.e., the chain-driven bicycle that we know today – could simplistically be attributed to a better design when compared to the earlier high-wheel bicycle. However, closer examination reveals that the obsolescence of high-wheel bicycles was due as much to changing values as it was to any technological superiority of its competitor; many users of high-wheelers valued them for their greater speed, while the safety bicycle presented a less dangerous alternative (presenting a broader appeal). The relative values of bicycle models were impacted on and altered by a wide range of social concerns and norms, and the changing fortunes of one model or another would aptly be attributed to an array of factors, not just "technical superiority." By judging technological successes and failures on the same terms, the real forces driving innovation and evolution may thus be more fruitfully identified.

Not Just a Theory

In addition to being a theory for conceptualizing technology, constructivism is also, and importantly, a methodology for examining what is exactly behind technological achievements and failures. This separates it from purely conceptual approaches like those of the previous chapters. Moreover, constructivism's ability to provide researchers with a concrete and principled method for approaching investigations of technology is undoubtedly a central factor for its enduring influence.

For the sake of brevity, we will omit the full methodology, and instead focus on what it gets right and where it is wanting.[9] In a nutshell, though, the constructivist analysis is divided into two stages. An individual technology is generally viewed as some sort of thing, some individual artifact. The first step in the analysis is therefore to conceptually break down a technology to its constituent parts and thoroughly study each one in isolation. In this stage, the constructivist looks to *where* component parts originate, from what disciplines, from what industries, and how the relevant social groups involved understand those component parts. After this, a researcher may move to the second step: examining how the component elements come together to form a single, coherent artifact. Because this second stage is predicated on the choices and actions of individuals, institutions, groups, etc., unsurprisingly, this stage is referred to as the "social construction element" of the method.

In any case, a fundamental point of the constructivist method is to constantly recognize that technology is never fully developed, even after it leaves the engineer's drawing board or the lab. Rather, technology is influenced by all succeeding actors who interact with it, including producers, distributors, users, etc. Users may even employ technology in ways that creators had not intended, in unforeseen and potentially unforeseeable ways – such as using social media apps to extort money – presenting a persistent need to re-examine the ways society constructs technology.

This makes the constructivist approach incredibly useful for understanding how technology changes over time. Focusing on the actual people involved in the design, production, distribution, and use of a technology also allows us to see not just the technical aspects of innovation, but the sociological side of technology as well. According to Pinch and Bijker, ultimately, technological development is a construct of the various humans that make up the ensemble of innovation, a stark contrast to the narrow "internal logic" view of innovation that we find in determinist thinking.

The merit of the constructivist position is highlighted when one considers that designers, in particular, have a large degree of control over how technology turns out, and as such, one arguably must view technology as at least partly constructed by the values and beliefs of those scientists and engineers involved in the design process. Their particular choices are moreover informed by their background values and beliefs, all of which are at least partly constructed by society. Thinking back to the example of Long Island's Parkways, these were designed to be low *only because* Robert Moses wanted them to be low. And he wanted them to be low *only because* his background values and beliefs created a reason for them to be low. Those values and beliefs were part and parcel of the society he grew up in and in which he lived, and they played an obvious role in the social construction of the Long Island Parkways.

Some Pitfalls

Despite the advantages explored above, constructivism is not without its shortcomings. First of all, the method, while presenting a useful heuristic for analysis, can present a pragmatic worry due to its potential demandingness. The approach tells us that in order to really understand technology, we must first break it down to its constituent parts, look at all of those parts, and all the people involved, and all the social structures that guide its development and use, etc. This entails an extremely demanding and complex task of actually figuring out where the borders of analysis are. Did I break down the technology as much as I could? Did I consider all of the people involved? Are some of the people and facts more influential than others? These are questions which may quite naturally arise, and delineating where to stop in any constructivist analysis is hard, if not impossible, to determine. Theoretically speaking, this could continue *ad infinitum*, making for a potentially comprehensive, yet practically impotent analysis.

Second, while casting a wider net than determinism, constructivism still presents an overly narrow understanding of technology and its development. Returning to Langdon Winner, though his analysis of the Long Island Parkways would seem to be presenting a view of technology, on the constructivist theory presented by Pinch and Bijker, that would not properly be the case. This is primarily because his account includes more than the constructivist would find appropriate. In fact, in the early 1990s, Winner criticized constructivism on the grounds that it is too restrictive and reductive (Winner 1993). For example, he argued that while constructivism views technology as resulting from the various actors and social systems in play, it gives no attention to the *actual* impacts that any given technology has after it has been deployed; constructivism simply looks at why a technology was successful (or not), an insufficient analysis overall. Similarly, constructivism rightly gives focus to the people involved in design and production of technology, yet it gives no space to the stakeholders who are impacted on by technologies and have no voice in the design process. This exclusion leads constructivism to an elitist understanding of why technologies are successful, ignoring those who are often most impacted by them.

Furthermore, constructivism avoids any normative stance concerning the potential benefits of other interpretations of technology. For example, Winner's interpretation of artifacts as the bearers of values ("politics" in his words) shows how technologies, as a function of their design, can have substantial impact on society, a point to which the constructivist cannot speak. In a related vein, the constructivist analysis generally ignores *framing effects*, i.e., the impact of context on a given technology or system (i.e., Goldin and Reck 2020). The most conspicuous examples of framing effects are opt-in or opt-out systems, like some pension plans. For a given plan, whether its default is that individuals are automatically in (and can easily opt out) or are automatically out (and can easily opt in) will massively affect the outcomes achieved. And this has nothing to do with the technology or system per se; pension plans are obviously socially constructed (i.e., the result of social pressures, workers' interests, etc.) and massively socially impactful, but whether a pension is opt-in or -out has nothing to do with the plan itself. Rather, this latter point is a simple design choice, not born out of society, but merely a function of some designer's choices, and this one choice directly and significantly impacts on both the technology and society.

Building on this, recent work on *nudging* shows how the design of a technology can directly influence people's behavior, and can be utilized to foster certain aims. Cass Sunstein and Richard Thaler argue that various things like buildings, instruments, and even other technologies can be designed with choice architectures that modify behavior without forbidding or overtly incentivizing any options (Sunstein and Thaler 2008, 6). Nudging thus leverages the work of behavioral sciences and economics to provide ways to alter behavior without compromising individual freedom or rational choice; a nudge can overcome a person's cognitive flaws, such as ignorance of a situation or bias, and direct them toward actions which are beneficial for them (Sunstein 2017).[10] Given these potentialities, constructivism's failure to account for the societal impact of technology is a substantial drawback.

Framing effects and nudges demonstrate how a value-free understanding of technology, like the constructivist view, fails to fully account for all of the nuances of technology. How a technology or system is framed or what nudges are employed inevitably involves value judgments, and implicates social values and norms that are deemed morally or politically preferable by decision architects. This shows that at least some technologies will not be value-free or neutral instruments (Prainsack 2020). This is even more clear (and worrisome) when we talk about cutting-edge digital technologies. In the digital sphere, tech corporations or private actors may fill the role of "choice architect," designing technologies with their own values, values that, in some cases, are not aligned with those of their users.

In sum, the constructivist approach certainly brings with it some helpful tools for deconstructing technologies' various parts and mechanisms, and highlights the role of people in technological development. However, constructivism is not without its issues, and a complete picture of technology must additionally be able to accommodate the impacts technology has on society and the ways technology (design) shapes the future.

Thinking with Fiction

Throughout this book, we've leaned on various works of fiction to help guide us in our understanding of technology. Fiction does not substantiate points, of course, but it does give us nuanced ways of understanding the world and the things in it. Before moving on to the next chapter, where we will present a more holistic and (hopefully) accurate account of technology, it merits coming back to a useful fictional analog to help us review what we have discussed thus far.

Thinking about Tolkien's *One Ring*, what would an instrumentalist say? We saw in Chapter 2 that some of Tolkien's own characters view it as a means to combat the very enemy that created it. In their eyes, its power is just as useful in their hands as it is in the hands of an enemy. It is a tool, nothing more. A technological determinist, however, would likely paint a picture of pessimism. After all, the *Ring*'s influence is so great, it is unlikely that anything can be done to stop its inevitable will to return to its master, and so all hope is lost. But where would a constructivist fit into our story? A constructivist would look to how the *Ring* was made, examining where and how Sauron forged it. They would likely look at the lesser rings as part and parcel of how the *One* fits into that larger ensemble. And when looking at Sauron's gifts to the other races, those lesser rings, we see a sort of rhetorical closure that is fundamental to the constructivist understanding of technological development. The rings give their bearers the strength and will to lead, and through that, they eliminate the need for those bearers to cultivate any strength or will to lead of their own. Without that need, those capacities are likely to wither, leaving the bearers without any alternative except to rely on their rings. This closure of alternatives is essential to how we understand the power of the *One Ring*, as its power is a function of its ability to control the lesser rings and their bearers. The closing off of alternatives is where the constructivist would locate the "power" of the *One Ring* rather than any sort of internal logic of the ring itself.

To close this chapter, it is worth asking whether all of these options for understanding the *One Ring* are mutually exclusive? Must we choose one? We have seen that each alone brings boons but also pitfalls, yet perhaps there is way to get the best of each. The next chapter will attempt to do just that, bringing us closer to a useful and authentic understanding of technology.

5

The Design Turn

> [T]he son of Cronos has put a greater worry in my heart that you, after too much wine, may start a fight amongst yourselves and then hurt each other, dishonoring your courtship and the feast. *Iron attracts a man all on its own.*
>
> Homer, *The Odyssey*

Our Dwellings Shape Us

On October 13, 2022, a Florida resident and his son received numerous alerts that their Ring doorbell camera had detected someone at their front door. Thinking that their home was being broken into, both armed themselves with handguns and went looking for the suspected intruder. As the pair emerged, the woman who had triggered the alerts saw them, and, thinking they were attempting to hijack her car, drove off. The father and son fired at her retreating vehicle, almost killing the woman. As it turns out, the woman was a neighbor who was dropping off prescription medication that was mistakenly delivered to her home (McDaniel 2022).

This may seem a fanciful and perhaps even ridiculous affair, but it is sadly not an isolated event. Karl

Townsend, a resident of Halewood, UK, received a Ring notification that his home was being broken into by four burglars (BBC 2022). Townsend armed himself with a kitchen knife and returned home, stabbing and killing one of the burglars in the ensuing altercation. Due to his actions, Townsend was sentenced to nineteen years in prison for manslaughter. And this has happened many more times, showing a growing problem (Greer 2022).

Thinking abstractly about these cases, we can ask ourselves what role the Ring doorbell plays in such incidents. More specifically, how does the design of the technology impact on these situations? If you asked an instrumentalist, they would likely say that the Ring camera is incidental to these cases, playing little, if any, role in what happened. After all, just as "guns don't kill people, people kill people," the Ring camera is nothing more than a tool, and how that tool is used depends entirely on the users of it. A technological determinist, however, might grant that the camera was partly to blame, but would add that this is a natural and unavoidable consequence of technologies like Ring cameras, and that a certain degree of social upheaval is to be expected with new devices. Such upheaval may also entail physical harm, but you #CantKillProgress. Social constructivists, on the other hand, would focus on the human actions of engineers, managers, executives, and users, seeing these as the "shapers" of the device (and what it does), rather than looking at the technical design itself.

Regardless of which interpretation one adopts, it is undeniable that technologies mediate and shape many of the ways we interact with one another. Moreover, despite their centrality to how we live our lives, we nonetheless have a tendency to see technologies as somehow separate from our non-technological "natural" world. Yet in reality, this is not the best way to view them. Modern technologies are built upon the scaffolding of previous systems, iterations, and artifacts that underpin and lay the foundation for those technologies we see today. Technologies like screws, wiring, and inverters permit the development of more advanced systems – like electric motors – requiring those foundational components. Thus, the technology of one generation serves as the groundwork for the next, and it is more appropriate to view each not as an isolated artifact, but rather as a system or infrastructure interwoven with other systems and infrastructures. Such systems and infrastructures are naturally impacted on by individual technologies, but also by society, and the end result of this is a view of technology as "socio-technical," a blending of technologies with the societies that created them.

Jeroen van den Hoven sums up this complex environment of "socio-technical-institutional" systems by drawing an analogy to Winston Churchill's comment that "we first shape our buildings and afterwards our buildings shape us" (Van den Hoven 2017). This position highlights that not only do our choices impact on our technologies, and vice versa, but also that the decisions made leading to a final design have real and potentially substantial impacts on both the technology developed and the society which plans to deploy it. To illustrate, think of how the SARS-CoV-2 pandemic has impacted the way people work and communicate with one other. The use of Zoom rose from ten million daily meeting participants in December 2019 to over 300 million in June 2020 (Iqbal 2022). And even today, both employers and employees continue to see the benefits of remote work. However, for all of its advantages, this makes people increasingly dependent on the hardware and software necessary for them to do their jobs and communicate with their colleagues. Thus, public health concerns drove a need for new technology (or greater adoption of existing technology), but the use of that technology has since shifted our general modes and organization of work, and across a broad spectrum of domains. As we change the ways we work in response to technology, technology is also apt to evolve, with one impacting the other in rounds of development. In the end, the freedom and ability to choose where and how we will work will almost certainly be a function of the sociotechnical infrastructures which have led to us having that choice, and how those infrastructures support or constrain our potential choices is apt to change over time.

Who's Responsible?

In a world marked by sociotechnical infrastructures, where every choice is shaped by and impacts on other systems, it may become difficult to navigate what it even means to be a free, individual agent capable of bearing responsibility for one's discrete actions. The complexity of our sociotechnical surroundings and their inextricability from our lives may even be thought to render traditional notions of individual responsibility meaningless (Scheffler 1995). What is needed, therefore, is a new way of thinking about morality, one that accounts for how technology intimately connects with our lives (Scheffler 1995, 228).

Prior to the twentieth century, communities were highly localized, but globalization and information technologies have massively expanded our ability to communicate, conduct business, and otherwise engage with the larger world. The ubiquity of smartphones and the internet furthermore allows us to achieve immense reach with our words and actions, and as such, our notions of responsibility must be amplified to account for that fact; traditional normative evaluations often focused on proximal effects and impacts close to home, but this changes when we recognize that a person with a smartphone can cause another to commit suicide at the opposite end the world (López-Meneses et al. 2020). Importantly though, technologies like smartphones are not isolated artifacts, but systems within larger systems. Sociotechnical infrastructures like telecommunications and the internet are cogs in the larger machine that make the phone in your hand so much more than *just* a phone. These systems boost the potential impact of any given individual beyond their immediate locality and onto the world stage, bringing deep questions on morality and responsibility.

Now, it is worth noting that some individuals have always been able to impact on the world stage, even during early portions of man's history. Moreover, any long-distance commercial activities required (and continue to require) the functioning of significant logistical and support structures. Thus, there is a sense in which sociotechnical infrastructures have always pulled humans into large interconnected systems of systems. Fundamentally, this interconnectedness of people, institutions, and infrastructures means that every person is capable of creating great change, and seemingly banal actions like ordering a particular product can implicate vast numbers of systems, institutions, and infrastructures.

In a high-tech world, the issue of responsibility becomes more challenging when we appreciate that the systems we utilize to carry out our actions are also mediating our actions. What this means is that rather than directly impacting on other individuals through our actions, we more and more impact on others by means of technology or by proxy of technology. This complicates our commonsense notion of responsibility by mixing (possibly inextricably) our responsibility with the responsibility of other individuals in the system, be they designers, controllers, or other users. Moreover, technologies do not emerge from nothing, but are the result of many design decisions made by a vast number of people across time. Taking this into account, Jeroen van den Hoven rightly argues that when thinking about moral responsibility, we should focus on

> (1) the *design histories* of the systems and technological milieu in which agency is staged and (2) the choices and role of *design agents* in the shaping of choice architectures. (Van den Hoven 2017, 19)

The takeaway here is that design, and those who carry out design, are central to understanding responsibility in what is fundamentally a designed world.

The Design Turn

As already discussed, when thinking about responsibility, we often intuitively think of cases involving individuals who are in some way proximal to each other. There may be mediating technologies, or minimal barriers, but all agents involved still somehow act on one another. However, as technologies become more advanced, it becomes less clear when someone is "proximate" to another, and the nature of sociotechnical systems makes any action one carries out likely to have interaction effects with many other agents' actions as well. All of this makes the attribution of responsibility difficult, to say the least. With this in mind, it becomes clear that, in a high-tech world, responsibility needs to focus less on pairs of proximal agents and more on the *design histories*, the *design agents*, and the *choice architectures* which support or constrain what people can and cannot do, and which make individuals more or less responsible. Put differently,

> The all-important new question prompted by the focus on the design is: who brought about these circumstances, who *made* them, who designed the context in which the morally problematic phenomena and hard choices are situated. In a fully designed environment asking primarily or exclusively about the obligations and responsibilities of an individual agent who finds him or herself in that situation may be barking up the wrong tree. (Van den Hoven 2017, 20)

By paying heed to how design and context affects individuals' choices, supporting or constraining particular (types of) actions, it becomes natural to shift our ethical thinking away from an emphasis on individual choice, and instead to a view of *ethics by design*. What this means is that we reframe our ethical theorizing to focus on how design choices shape the contexts in which individual choices are made, and how the options available to individuals are expanded or limited by prior design choices. We can already see this shift to a design-oriented ethics by looking to recent cases like the Dutch rail operator Prorail altering their approach to mitigating rail accidents. Previously, such accidents were generally viewed primarily with an eye to the potential human error introduced by the train's driver. However, there is a growing understanding that there are contextual factors impacting on a train driver's ability to recognize stop signals or respond in a timely fashion, and these are the result of previous design decisions over which the driver has no control. By examining the entire design and use history leading up to an accident, it is therefore more likely that accidents may be mitigated (Van den Hoven 2017, 21).

There are many cases like this, and all betray the complexity of modern systems, institutions, and infrastructures, all of which can challenge our ability to adequately determine responsibility or explore ethical issues using our conventional person-centric methods.

Indeed, we are apt to find that there are in fact many people, institutions, and changing social situations which contribute to any given context and option set we are presented with today, and these various factors may also span over time periods, further complicating matters. This, of course, does not remove responsibility, but it does show how nuanced assignments of responsibility must be, and moreover highlights that they must account for both the choices an individual makes in a particular situation, as well as the (design) choices of other individuals that led to that situation in the first place. By thinking along these lines, we may do justice to all individuals involved. This is what van den Hoven calls the *Design Turn in Applied Ethics*, and it has led to a broad field which aims to harness design to make a material impact in the world (Van den Hoven 2017, 23).

Choices, Nudges, and Architectures

Thinking back to the example of the Long Island Parkways, we can see how technologies, as a function of their design, can limit the choices available to individuals. In the case of the parkways, they closed off alternatives to those reliant on public transit while preserving options for those with automobiles. And we can see this opening and closing of alternatives with almost every technology, rudimentary or advanced. A screwdriver opens new construction possibilities for stronger and more durable structures, but by its design, limits the array of items which may be used with it; i.e., it requires screws with particular head types. Moreover, a screwdriver opens possibilities for using screws, but it is an incredibly poor tool for driving nails (effectively removing that option). Smartphones provide a more high-tech example, where the interface permits and extends certain actions like communicating, downloading and running applications, or taking photos. However, that same interface, by its design, limits one to installing only applications compatible with the operating system, makes it more difficult for users to limit their availability, and can even make it so that users are incapable of making purchases or carrying out everyday tasks (in cases, say, where a phone is used to pay for items, but its battery has run out).

Whatever the technology in question, the point is that they shape the contexts in which we make choices, and they open and close alternatives to us. As such, technologies have a material and substantive impact on society, as their design may foster or hinder certain courses of action. Some design choices allow for certain behaviors or make them more probable. Recalling the example of nudging from the previous chapter, this is a clear case of how design can be used to influence the likelihood of certain behaviors rather than others, and this may be employed for the most banal of purposes. For example, a fly-shaped sticker placed in a urinal makes it more likely for people to aim their stream in such a way that messes are minimized (Evans-Pritchard 2013). Likewise, a transparent trashcan with your favorite sports teams on it is more likely to have you "vote" with your trash, rather than throw it on the floor (Peters 2015). What this means is that engineers and designers are best understood as *choice architects*, whose decisions in design foster or constrain the likelihood of certain behaviors emerging (Van den Hoven 2017, 24; Thaler et al. 2012).

Interactionalism

There are numerous ways to put the insights of the design-oriented approach to ethics into practice, and we will explore these in detail in Chapter 7. For now, though, it is important that we first get a clear view of the design turn's characterization of technology, what we will call *interactionalism*.

Technologies are central to how we live, structure our environment, experience the world, communicate, love, eat, and even sleep. The cars we drive, the smartphones in our pockets, and the apps we use to find romantic partners and order food are just some of the daily systems that have become almost mundane, yet at one point in time would have been unthinkable. There are no clear or distinct boundaries that separate human society from the systems impacting on our lives and society, and this forms the foundation of interactionalism; society and technological systems must be viewed as one, as *sociotechnica*l systems. This blending of society and technology reflects that interaction effects can move both ways as well, with society altering the development space for new technologies and new technologies pushing society to reevaluate its norms, regular practices, and even goals. This way of thinking shows that it is inaccurate to view technologies as simply tools in a long line of increasing development. Instead, they form a part of the social world, and are in constant flux due to shifts in the societies using them.

The core of interactionalism is the conception of technologies as interactional, meaning that societies, individuals, institutions, and organizations shape the systems they design and deploy, and, in turn, those systems shape their shapers (Friedman and Khan 2003). This leads us to view technologies as, at least in part, inherently good, bad, or neutral (pace the instrumentalist). This is because technologies will be designed with the values of their designers, will (partly) embody those values, and the values themselves may be judged to be good, bad, or neutral. Moreover, systems may be intentionally designed to embody particular values (like

Moses' parkways), but even when that is not the case, a designer's values will usually impact on the system in important and significant ways (Kranzberg 1986, 545; Friedman and Hendry 2019, 29). What a technology "does" then is a consequence of how its features were designed, and this is contingent on the aims and underlying values the designer had in mind. Dykes are designed to increase the safety of coastal regions, the vertical forest condominium building in Milan embodies sustainable living in urban environments, and word processing platforms have built-in accessibility features to enable their use by individuals with visual impairment. All of these are examples of designs, of particular technologies, but each clearly exemplifies particular goals and values of their designers.

None of this entails that designers have full control over all the values embodied by a given technology. Rather, the complex landscapes of use for technologies means that systems can bring about outcomes which designers did not foresee, or possibly could not foresee. Yet even so, there is a space in which designers can meaningfully and significantly shape their technologies to foster certain values. Recalling the example of pension plans or group insurance policies from the previous chapter, whether or not such systems are designed to frame user choices by an opt-in or opt-out approach can have a large impact on outcomes (Orlikowski 2000). Moreover, these framing effects can be consciously harnessed by designers in order to promote certain outcomes. This shows not just the importance that design decisions can have for behavior, but also the very real possibility of engaging in social engineering via design. Large corporations like Google and IBM are well aware of this, and by exploiting these nuanced realities of our sociotechnical world they can engineer markets favorable to their products and manufacture consent for their economically charged invasions into our personal lives (Tenner 2006; Gabberty and Vambery 2008).

Importantly, whatever technologies develop, they will never do so independently of designers, and this creates a crucial first introduction of values into the design process. When a firm investigates the possibility of creating some new technology, designers may, of course, conclude that it should not be developed in the first place. In that way, values also impact on whether technologies get to the drawing board at all. For example, some designers and policymakers may seek techno-fixes to address current problems we are facing; climate engineering technologies like solar radiation management have been theorized as a potentially powerful way to fight global climate change. However, some engineers and scientists, fearing the risk of unforeseen catastrophic side effects of such technologies, have argued that it is better to focus on policy mechanisms rather than untested techno-fixes that may well cause more harm than good (Wolff 2020). Another timely example is digital contact tracing systems developed to help combat the SARS-CoV-2 pandemic. Some governments (e.g., Sweden), were, however, concerned about the potential privacy violations involved with developing and deploying such a technology, and decided to scrap the proposed national platform (Carlsson 2020).

Interactionalism also looks at how a single technology can have different effects – sometimes unforeseen or even unforeseeable – depending on the different contexts in which it is used and which people use it. With this in mind, interactionalism implores designers to remember that some unforeseen impacts may be unwanted and perhaps even dangerous, a point arguing for caution. Moreover, a technology's effects can stretch beyond the particular time and place where the technology is used and can have widespread, even global, impacts. Thus, the simple act of creating a technology brings with it a host of (potentially global) issues that may emerge over long spans of time. And this brings us to the more general aspect of time, something key to understanding the interactional view of technology.

The impacts of technologies are not only immediate, but may sometimes take years, decades, or even whole generations to manifest themselves. The dynamic, changing, and complex web of relationships we have with each other, our technologies, institutions, and infrastructures, moreover, means that by the time a technology has caused its more remote effects, our values may have changed (compare with De la Bellacasa 2012, 2017). For this reason, it is important for interactionalism that technologies be designed to be modular, reflecting the changing values that different peoples have across time. On the societal side, interactionalism also argues for the education of people concerning the changing impacts a technology may have, teaching them ways to modify or work around a technology's unwanted behavior(s).

Fundamentally, the interactional stance rejects the deterministic outlooks of technological determinism and social constructivism and the value-neutrality of instrumentalism. Each of these traditional views have something important to say, but their unary conceptions bring problems as well, and the interactional stance aims to gather their goods while tempering their shortcomings. At its core, interactionalism places faith in humans to direct design well but provides a guide so that we do not make the mistakes of Dr. Frankenstein.

The interactional stance is born out in countless real-world examples, but we can see it even in fiction as well. Returning to Tolkien's *One Ring*, its interactional nature is laid bare by the fact that it can be used as an instrument of power, but it does so *as a consequence* of the values of its designer. Moreover, it is inextricably connected to other devices and peoples (i.e., the lesser rings and their bearers), and whatever is done to the ring will impact on those other devices (technologies) and peoples (societies) as well. In short, the One Ring was created as an extension of Sauron, and looking at the *One Ring* as a simple circle of metal would be a life-threatening mistake.[11]

Sauron and Robert Moses are similar in that they were both keenly aware that the design decisions they were making would have impacts across society and time. Unfortunately, cases like Moses' parkways exemplify how those with bad intentions can appropriate design to further ignoble values. Yet, even so, the impacts of a technology can change, just as the parkways have changed, and this provides a hopeful outlook; we can design technologies for good, and even bad technologies may lose their potential to do evil in the world. In any case, designers and engineers have the power to create embodiments of value in the world, and with that power comes a responsibility to design for good ends, and with an eye to the changing nature of technologies and values.

6

Responsible Innovation

> Determine that the thing can and shall be done and then we shall find the way.
>
> Abraham Lincoln[12]

The first part of this book examined the "what" of technology. What is technology, in what ways has it traditionally been conceived, and what benefits and pitfalls attend those classic perspectives? By way of instrumentalism, technological determinism, and social constructivism, we eventually came to what scholars currently view as the most authentic way of understanding technology, namely *interactionalism*. Now, this is all well and good, but even accepting this to be the case, there is still the crucial question of *how* we ensure that technological innovation tracks the core findings of *interactionalism*? More specifically, how can we ensure that our technologies, which both shape us and are shaped by us, promote our core human values?

This chapter, and those that follow, will move away from the *what* and instead focus on these questions of *how*. How can we, taking the interactional stance seriously, begin to design with it in mind? How can institutions and research-funding bodies promote responsibility in design? How can we normalize interactionalism in our broader discussions of technology in media and popular culture? These questions (and many others) have guided both research and funding allocation over the last ten years, and many of the underlying precepts of the interactional stance can be brought under the umbrella of what government bodies have come to call *responsible research and innovation* (RRI), or, more shortly, responsible innovation (Von Schomberg 2013; European Commission 2013).

Responsible innovation emerged mainly from concerns over the ethical, social, legal, and environmental impacts that technologies (can) have,[13] and was given further impetus from research in science and technology studies, policy and governance studies, and a variety of approaches to designing for values (which we will see in the next chapter). One core premise of responsible research and innovation is that technologies have real and material impacts on society, and bring with them a host of ethical (among other) issues which can be exacerbated or mitigated by the design of those systems (Smallman 2018). Because of this, according to RRI, the analyses and assessments of technological impacts are a central and necessary focus of any innovation program.[14]

Embodied in Consumer Movements

A core idea of the *interactional* stance is that innovation and social spaces interact in fundamental ways. This means that systems may (or are likely to) have both positive and negative consequences for society. A quick look reveals the extent to which innovation has impacted on various areas of our lives, from housing to shopping, employment to entertainment. Moreover, failure to bring innovation (practices) in line with social concerns can be detrimental to companies as well; Apple has continually been accused of allowing for very poor working conditions in its factories in China, damaging its public image (and potentially its profits) (Albergotti 2019). Thus, interactions move in both directions, and numerous companies and recognizable brands have modified their innovation tactics in response to interactions with social spaces.

However, others have adopted a different approach, proactively innovating strategies which address social problems at the outset. For example, the sock company Bombas made the bold decision to develop a radical business strategy around a "One Purchased, One Donated" objective. To summarize their concept, for every pair of socks that is purchased from the company, Bombas donates one pair to a person facing homelessness (Bombas n.d.). Adopting a responsible innovation approach, Bombas founders Randy Goldberg and David Heath aim to alleviate social concerns through market forces, thus contributing to a broad positive impact on society.

The Bombas mission was sparked by a social need that the founders saw as requiring addressing, and they designed a means for doing so. However, not all problems will be noticed by heads of corporations themselves. Thus, directly listening to those impacted by certain technologies, i.e., stakeholders, provides a method for innovators to locate potential concerns and begin to develop novel ways to address those concerns. This can allow consumers to push for technologies that align with their ethical values, incentivizing companies to innovate according to those values. For example, Fairphone, an Amsterdam-based electronics manufacturer, has created a smartphone that aims "to last," thereby reducing waste. Moreover, the company sources both recycled and responsibly mined materials, to ensure not only that environmental and labor concerns are respected, but also to increase industry and consumer awareness of the possibilities for providing high-end technological innovations while making a positive impact (Fairphone n.d.).

Talking with Stakeholders

Advanced technology, which is nearly ubiquitous around the globe, can contribute greatly to societal progress and human flourishing, but can also be a force for oppression or used to entrench long-standing injustices. Gadgets, infrastructures, pharmaceuticals, and logistics systems all have the ability to cut both ways, and both popular media and academia are replete with examples of ethical issues emerging from things like artificial intelligence, social networking platforms, biotechnology (like CRISPR), digital voting machines, or space colonization programs (to name just a few). Responsible innovation takes these issues seriously, aiming to find practical and actionable solutions, both through engineering and public policy (as these impinge on one other). And since the issues being addressed are, first and foremost, born of our values, it is crucial that responsible innovation take stock of these.

To ensure that values are tracked when designing technologies, responsible innovation requires, first and foremost, stakeholder participation. Input and feedback from professionals, policymakers, consumers, and non-governmental organizations provide the foundation for guiding design toward socially beneficial ends. And here the interactional stance plays an important role. Technology, when understood *interactionally*, can be characterized as "systems of socio-technical systems." This means that a system being developed is impacted on and impacts on the institutions, infrastructures, people, and groups that form the broader sociotechnical system (Van den Hoven 2014, 6). Thus, interested and impacted parties must participate in technological discussions, as their views form part and parcel of the resulting technologies. For this reason, stakeholder engagement is a common element among various responsible innovation programs.

When to Responsibly Innovate

A further important question for responsible innovation is *when* this should occur. In popular culture, and even in much of the academic literature, the questions being asked are often how technology *could have* been designed more responsibly. More specifically, much of the literature looks at already existing technologies, and often only when something goes wrong, when there is an accident, or when a system displays unpredictable behaviors. Thus, the default attitude is to look to responsible innovation *after* a technology has been developed, produced, and implemented in society. There may be examinations of ways to alleviate issues which arise, but unfortunately, these often come too late and make too little difference to those who have taken the brunt of a technology's impact(s). This tendency to wait until problems come up is worrisome, given that we often will not see the full consequences of a technology until it has been deployed in society and, in many cases, become pervasive, normalized, and potentially difficult to remove. In such cases, it can be challenging, if not impossible, to intervene and address emergent problems.[15]

Values and Design

So responsible innovation demands that human values, in particular, stakeholder values, be included not just after technologies fail, but also during the design process, and indeed before any designs have been drawn up. However, as we will see in the next chapter when discussing "value sensitive design," it is also important to stress that human values are not add-ons or secondary features to be implemented *ad hoc* in technologies, but are design *requirements* similar to other technical requirements a system must fulfill in order to function correctly (Van den Hoven 2007). This shows the importance that design has and the responsibility that designers have in how they approach their work. Moreover, this means that for responsible innovation to take place at all, we need to be able to articulate what our values are, what presumptions our values make, and what design criteria our values imply.

As an example of the ethical and social dimensions of innovation, take the production and distribution of cheap smartphones to poorer individuals and groups. On the one hand, one might imagine that increased access to technology and the internet might decrease the digital, economic, and social divides between developed and developing countries. However, according to a 2017 study, the most vulnerable smartphone users are those whose devices are susceptible to fraud and harassment, and more affordable, lower-end phones are inherently less secure than their more expensive rivals (Maréchal 2017). Moreover, poorer individuals will replace their devices less often, and older models often have known vulnerabilities, limits in their ability to handle software upgrades or, if they can be upgraded, often do so at the cost of basic performance.[16] Thus, while providing affordable smartphones to developing nations and vulnerable groups would in theory be an objective worth pursuing, in practice it can lead to as many problems as it solves. This is especially problematic given that those opting for these phones are generally from lower socioeconomic backgrounds and are thus likely to rely on them as their sole means of internet access.

The problem with cheap smartphones was that, though they could clearly bring a host of economic, social, and human benefits to those using them, privacy and security concerns with the design, distribution, and long-term use of such phones were not sufficiently considered. As a result, there were no plans or even intentions to make those devices more agreeable or beneficial for already vulnerable groups. These issues might not have emerged if engineers had taken the privacy and security of these smartphones more seriously from the beginning, and worked harder to build privacy into the architecture of the systems by, for example, simply creating a hardware platform that could more readily accept subsequent security patches.

Compare this to *privacy-enhancing technologies* like virtual private networks (VPNs). VPNs were designed to not only support moral values (like privacy) but also increase economic benefits (e.g., by increasing connection speeds and enabling smoother file-sharing). VPNs prevent users' privacy from being infringed upon by internet providers or governments (Deng et al. 2011) and, along with other privacy-enhancing technologies, are used by electoral bodies to ensure that users can safely and securely connect to the internet and transmit confidential data (Beroggi 2008). These protect citizens' private data, ensure election integrity, and even present a means to access the internet to those who might otherwise be blocked behind regional restrictions.

Moral Overload

What we can learn from the case of affordable phones, as well as the use of privacy-enhancing technologies like VPNs, is that values (like privacy) must be accounted for early on in the design process, and must be seen as *requirements* akin to functional concerns like affordability, performance, or connectivity. Responsibly designing an affordable smartphone would thus entail both anticipating the various problems which might emerge over the lifespan of its use, and making targeted design decisions to balance technical requirements and economic values with moral requirements and human values. In the case of affordable smartphones – those most likely to be purchased by vulnerable groups – designers had at their disposal a host of options for balancing concerns like affordability/cost and long-term respect for users' privacy. More strongly, the design of affordable smartphones was not predetermined, but rather the result of designers' choices, and hence, privacy *could have* been designed for (Garcia and Jacobs 2011).

With this in mind, the natural question which arises when designing any new technology is which values should be considered and incorporated. Unfortunately, there is no easy, straightforward, or universally acceptable answer to this question, and this is a fact which we will continue to (have to) deal with and which cannot be resolved completely. That being said, many projects explicitly employing some responsible innovation approach have developed tools to help us check whether our inventions map to our values. This is a hopeful sign.

Yet even so, sometimes we will find ourselves in situations where we have too many value commitments and cannot respect them all. This is a situation referred to as *moral overload* (Van den Hoven et al. 2012). Though moral overload presents a potentially serious issue, design can offer some potential solutions: the demands of urban density and concerns of environmental sustainability can lead to creative solutions like the Bosco Verticale (vertical forest) in Milan; doubts about election integrity can be calmed while simultaneously increasing efficiency and accessibility by using onion routing, the secret ballot, and VPNs (Namara et al. 2020; Fortier and Ornstein 2002); and the desire to get a good quiet night's sleep without impinging on safety can be seen in things like the Bose Sleepbuds II (Bose n.d.). At this point, one might be tempted to argue that these are clear examples of techno-fix solutions (hearkening back to Chapter 2). However, responsible innovation is not only about the design of technology but also the social spheres in which innovation is born, grows, and changes. In some cases – as we shall see with value sensitive design in the next chapter – after thoroughly investigating the potential impacts of a system, designers may choose that the best course of action is to not create it at all. Thus, while change and innovation is nearly always a possibility, we may echo Aristotle and leave open whether some particular change is necessary (D'Angour 2021, 84–85). Sometimes the best course of action may be to do nothing at all,[17] a point which can be well-defended on the responsible innovation approach.

7

Approaches to Ethics by Design

Designing with Ethics for Ethics

Technology impacts heavily on our lives, and moral questions about its use and control have been debated for centuries.[18] Recently, the focus has shifted to how we can create practical frameworks for ensuring that moral considerations are given appropriate attention early on in innovation programs, when they still have a greater potential to have an impact. Responsible innovation provides such practical frameworks, drawing on a range of techniques and lessons rooted in many different traditions, and allowing for implementation in a variety of ways. To fully grasp the potential of responsible innovation, we must adopt an interdisciplinary approach that transcends the boundaries of STEM. Insights from the social sciences, including ethics, sociology, and economics, are indispensable. They provide context and understanding of the societal impacts that raw technological advancements may overlook.

Over the past few decades, several approaches to design have been created and adopted, and in this chapter, we will take a brief look at some of the more popular ones, such as *universal design*, *inclusive design*, *participatory design*, *human-centered design*, and, finally, *value sensitive design*. we will also look at how and why the latter, value sensitive design, has come to dominate the responsible innovation literature.

Universal Design

Culminating in what was to be called The Center for Universal Design at North Carolina State University, founder Ron Mace formalized the approach of Universal Design in the late 1990s (CEUD n.d.). *Universal Design* is a concept in design and architecture that aims to create buildings, products, and even whole environments that are accessible to everyone, regardless of age, socio-economic class, or ability (Mace 1990). The approach is also sometimes known as *inclusive design,* which has garnered it increased attention and importance in recent years as society has placed more emphasis on diversity and inclusion (Persson et al. 2015). There are a variety of benefits of Universal Design, including increased accessibility and usability for people with different degrees of abilities, increased safety and convenience for everyone, and the potential to save time and money by designing products and spaces that can be used by a larger number of people. Not only this, but the approach can be used to support social inclusion by designing systems in such a way that everyone has access to the environments we build and the products and services that are available within them.

There are a number of examples of Universal Design that we can find in everyday life, such as ramps and elevators in public buildings, which help to make them accessible to people in wheelchairs or the use of audio and visual systems to aid people with visual or hearing difficulties in safely crossing roads. These are just some common examples that illustrate how the approach can help to make products, environments, and systems more accessible and usable for people with different capabilities.

However, while Universal Design has many potential benefits, it also brings drawbacks and challenges. One of the central difficulties of the approach is that it can be expensive to implement; in order to make buildings and environments accessible to everyone, designers and architects often need to incorporate tailored and bespoke features and technologies, such as ramps, elevators, and wider doorways, all of which can add to the cost of projects. This can present a barrier for those organizations and individuals which lack the funds necessary to incorporate universally inclusive features into their designs.

Beyond this challenge of universal design, it can be difficult to find the right balance between the needs of different users. For instance, a ramp meant for a wheelchair user might not be appropriate for a blind person, and a design that is simple to use for an older person might not be clear to a younger person. As a result, it can be challenging for designers to create environments that are truly *universal* and which meet the needs of everyone. Likewise, some critics argue that universal design can lead to a lack of individuality and creativity in design (Patrick and Hollenbeck 2021); because universal design focuses on accessibility and practicality, it may limit artistic expression or creativity. This may be viewed as a restriction by designers who seek to produce distinctive and avant-garde surroundings.

Participatory Design

Participatory Design is an approach that involves actively involving many potential stakeholders in the design process, be they users or other members of the community who may be impacted on by a technology. Participatory design seeks to incorporate the diverse perspectives, needs, and experiences of these individuals in the design of products, services, and environments. The origins of this approach can be traced back to designers and researchers in Scandinavia in the 1960s and 1970s (Cumbo and Selwyn 2022; Asaro 2000), and since then, it has evolved and seen application in a variety of fields such as human-computer interaction, urban planning, and public health.

Increased user happiness and involvement, better design solutions that cater to users' wants and preferences, and a greater sense of ownership and empowerment among stakeholders are just a few advantages that can result from Participatory Design (Thinyane et al. 2018). Additionally, the approach can encourage stakeholder cooperation and co-creation, resulting in more original and imaginative design solutions. For example, Carolin Mees describes how Participatory Design can be used in the development of a community garden in a low-income neighborhood. In this case, the approach might involve working with members of the community to identify a suitable location for the garden, soliciting input on the types of plants and features they would like to see, and involving community members in the planning and construction of the garden. The end result would be a garden that is tailored to the needs and preferences of the community, and that reflects their involvement in and ownership of the design process (Mees 2018).

In a broader sense, Participatory Design involves bringing users, designers, and decision-makers together to build a system, service, or product. Users are given the chance to actively participate in the design process in these initiatives and are given several opportunities to offer suggestions and feedback. For instance, the users can be requested to evaluate prototypes, offer comments on design ideas, or take part in co-creation sessions with the designers (Jones 2018).

This approach is also not without its limits though. One drawback of Participatory Design is that it can be time-consuming and resource intensive. Involving multiple stakeholders in the design process can require significant coordination and communication, and can slow progress. Additionally, not all stakeholders may possess the knowledge or experience necessary to actively contribute to the design process, which might result in less effective or efficient design solutions than they otherwise might be. Reaching an agreement on a design solution can also be difficult in instances when stakeholders have divergent interests or goals, making Participatory Design challenging in such circumstances (Bannon and Ehn 2012). Moreover, the approach can be less effective if users do not themselves have a clear understanding of their needs or preferences, or if users are unable to effectively communicate their ideas to designers (Van der Velden and Mörtberg 2015).

Human-Centered Design

The concept of *Human-Centered Design*[19] (also known as *user-centered design*) has its roots in the mid twentieth century, in the fields of psychology and engineering (Ranftler 2016), but has since gained widespread adoption across design disciplines. It is now widely used as a framework for creating products, services, and systems that are intuitive, usable, and accessible to all users.

The requirements, preferences, and skills of the users of a good, service, or system are the main focus underlying the design philosophy of *Human-Centered Design*. The approach has a wide range of advantages, including the potential to produce fresh and creative design solutions, improved usability and accessibility, and increased user happiness and engagement. By keeping users at the center of the design process, Human-Centered Design can help to ensure that final designs are tailored to the needs of the intended audience, and are easy and intuitive to use, leading to better user experiences. Human-Centered Design's focus on usability and the human side of technology can also reduce the need for costly training or support. Additionally, by making sure that designs are useful and accessible to people with a variety of abilities, this strategy helps promote social inclusion and equity (Holeman and Kane 2020).

In projects that employ Human-Centered Design, designers start by gathering data on users, such as demographic information, the goals and needs of users, and their abilities and constraints. The designers use this knowledge to develop user-friendly, intuitive design concepts that are specific to those users. For instance, a Human-Centered Design project might entail developing a smartphone app that is simple to use, offers users the features and information they require, and is accessible to users with a variety of skills (Nimmanterdwong et al. 2022).

Similar to Participatory Design, this method involves acquiring and analyzing information on the needs, preferences, and skills of users. This can be a difficult process that takes much work, and it may call for specialist knowledge or sophisticated research techniques. Also, like Participatory Design, when users do not have a clear understanding of their own needs or preferences or when they are unable to effectively communicate their ideas to designers, this can greatly hinder Human-Centered Design efforts (Steen 2012). In these cases, it can be difficult for designers to meet the needs of users. Finally, when users lack a strong voice or influence in the design process, such as when they are children, older individuals, or persons with impairments, Human-Centered Design may not be the appropriate method. Other design strategies might thus be more useful in certain circumstances to guarantee that the intended audience can use and access the final product (Guffroy et al. 2017).

Value Sensitive Design

There are a host of examples demonstrating how values are embodied in technologies, and with this in mind, scholars like Batya Friedman and David Hendry formulated a novel approach to design that pulls from the other approaches so far discussed in order to design technologies explicitly *for* human values (Friedman and Hendry 2019). This approach, called *value sensitive design* (VSD), argues that the design of a technology impacts "directly and systematically on the realization, or suppression, of particular configurations of social, ethical, and political values" (Flanagan et al. 2008).

Friedman and others created VSD by expanding on ideas that emerged from the human-computer interaction community to highlight the social and moral aspects of design (Friedman 1996). In VSD, the goal is to connect those involved in system design with those who are directly or indirectly impacted on by the system (i.e., the direct and indirect stakeholders, respectively) while infusing the design process with human and moral values. This approach to design focuses on understanding the values and beliefs of users and other stakeholders, and on designing solutions that align with these values. Value sensitive design is concerned with ensuring that both the design process and final designs are fair, equitable, and respectful of the needs and preferences of all stakeholders. Some of the key principles of the approach include considering the diversity of stakeholders and their values, promoting transparency and accountability in the design process, and ensuring that the design is responsive to the changing needs and preferences of users over time, even across generations.

By considering the values and beliefs of all users and other stakeholders, VSD can help to prevent systems from causing harm or injustice to any group of people. For instance, the Fairness in Design (FID) framework, a VSD approach, concerning the development of AI systems, permits surfacing and addressing complex fairness-related issues through a game-like approach involving prompt cards. This approach, as demonstrated in real-world case studies, not only aids in identifying potential concerns from various stakeholder perspectives but also fosters a more inclusive design process that promotes collaboration and co-creation among diverse team members, leading to more innovative and creative design solutions (Zhang et al. 2022). Here, VSD demonstrates its ability to be effective in guiding AI software designers to prioritize fairness in their design decisions, underscoring the applicability of VSD principles in contemporary technology development. Moreover, it emphasizes the importance of transparency and accountability, providing a tangible example of how VSD can ensure that technologies are sustainable and adaptable over time, particularly in addressing the ethical dimensions of AI systems. Such practical applications of VSD demonstrate its potential in creating technology that is not only functionally effective but also ethically attuned to the values of a broad array of stakeholders.

Beyond this, one notable aspect of VSD is its explicit orientation toward thinking about technologies across lifespans and even generations. This is first and foremost because technologies can have multi-generational impacts (like Moses' bridges), but the approach also recognizes that people's needs and abilities can change over time and that the design of products and environments should be flexible and adaptable to accommodate these changes (Yoo et al. 2016; Umbrello 2022b). In what VSD calls a multi-lifespan co-design project, designers work with users of different ages and abilities to create design solutions that are usable and accessible for all, and which reach into the future. For example, a multi-lifespan co-design project might involve creating a public park that is accessible and enjoyable for children, teenagers, adults, and senior citizens. By considering the needs and preferences of people of all ages and abilities, multi-lifespan co-design can help to create design solutions that are inclusive, equitable, and sustainable over time.

Like the previous approaches, value sensitive design can be time-consuming and resource-intensive; gathering and analyzing data on the values and beliefs of users and other stakeholders can be a complex and labor-intensive process, and may require specialized expertise or advanced research methods. Additionally, involving stakeholders in the design process can require significant coordination and communication, which can slow down the design process. Additionally, VSD may not be appropriate in situations where the design process is highly constrained by technical, legal, or other factors, as such situations may preclude the accommodation of stakeholders' values and beliefs. In these cases, other design approaches may be more effective in ensuring that the final design meets the goals of the project.

Still, one major advantage that VSD has is that it presents unique advantages for policymakers. It establishes a design framework that embodies ethical, social, and political considerations right from the conception of a technology, like AI. For instance, consider the example of ChatGPT, a cutting-edge language model developed by OpenAI. Technology that can generate text that sounds like human speech, such as ChatGPT, is unquestionably transformational. The social repercussions span the banal to the deep, affecting industries like entertainment, customer service, and education. As a result, it was impossible to separate the design process from any potential societal implications. Traditional technology policymaking, however, frequently finds it difficult to keep up with the rapid speed of technological progress.

VSD, by putting values at the forefront, enables the policy to co-develop alongside the technology. Instead of being a reactionary measure, policy becomes a proactive component of the design process (Umbrello 2021). In the case of systems like ChatGPT, this could mean involving stakeholders such as educators, linguists, psychologists, ethicists, and AI specialists early in the design process to anticipate potential societal impacts. Rather than finding ourselves in situations where calls for complete moratoriums are promoted because of emergent issues that are hard to stop (Knight and Dave 2023). Additionally, VSD promotes accountability and openness in the design process, both of which are crucial for developing trust between stakeholders and designers. Transparency in AI systems like ChatGPT can entail providing details about the language model's training, the data used, and the measures used to reduce bias. This openness aids in the development of more sensible and practical regulations by decision-makers. Importantly, VSD's multi-lifespan approach ensures the adaptability of policy over time. Just as technology evolves, so too should the policy that governs it. Policymakers, through the VSD lens, can anticipate and plan for the changes that a technology like ChatGPT may undergo over time.

The modularity, comprehensiveness, and constant improvement of VSD over its thirty-year history make it a strong candidate for how to actually implement responsible innovation. It is likely that different design approaches will be appropriate in different situations, and the best approach to use will depend on the specific goals and constraints of the design project.[20] It is also in general important to consider the strengths and limitations of each approach and to choose that which is best suited to the specific goals and constraints of a particular project. However, VSD's inclusion of many elements of the previous approaches makes it a strong *default* for innovating responsibly.

Values and Preferences

One of the primary features distinguishing value sensitive design from Universal Design, Participatory Design, and Human-Centered Design is that the former makes an explicit commitment to moral values, rather than the mere preferences and needs of involved stakeholders. This is particularly important for technology ethics more broadly, given that technologies impact not just on our preferences – which are dynamic and often changing – but also on our values. The inclusion, or better yet, the centralization of values, is thus particularly important for good design to take place.

This ability to take moral commitments into account is one of the primary advantages VSD has over its competitors, giving the approach a more objective foundation than what can be had by looking only to preferences (which are subjective and change from person to person). In particular, the founders of VSD argue that at least three fundamental values are universal: *human well-being*, *justice*, and *dignity* (Friedman and Hendry 2019, 173). These values then guide further exploration of human values, and serve as a baseline by which design processes may be judged.[21]

This approach to framing moral values furthermore allows designers to have a hierarchy of values, granting a clearer way to solve cases of moral overload (i.e., where designers are confronted with too many moral commitments, some of which conflict). Thus, designers can look to the value hierarchy and prioritize the most important or urgent moral decisions, and focusing on those decisions first, think about how practically to meet the moral commitments of stakeholders who may be impacted on by the technology being designed. This allows for creative and often unintuitive design architectures to emerge, allowing for more moral values to be incorporated than might at first seem possible. Because of this, VSD presents an optimistic outlook which allows for abundance, growth, and well-being to be pursued creatively while also focusing on other concerns like global equality or environmental protection, and does so without requiring us to abandon progress or necessarily scale back economic developments. By involving users in the design process, value sensitive design also fosters an inclusive environment which not only respects users' autonomy and agency, but also empowers them by giving them a voice and say in the design of those products and systems which affect their lives. This inclusion moreover helps VSD reach design solutions that are more responsive to the needs, preferences, and values of users.

The underlying philosophy of VSD is to, in the first place, avoid creating situations where technology can become a negative and dominant force in our lives, and in this, it aligns directly with responsible innovation and technology ethics more broadly. Recalling our example of the *One Ring*, Isildur's refusal to destroy the Ring, despite the counsel of Elrond and Círdan, stemmed from his view of the Ring as recompense, a beautiful object "fair to look on," too precious to be discarded, not recognizing its intrinsic purpose to dominate:

> For Isildur would not surrender it to Elrond and Círdan who stood by. They counselled him to cast it into the

> fire of Orodruin nigh at hand... But Isildur refused this counsel, saying: "This I will have as weregild for my father's death, and my brother's. Was it not I that dealt the Enemy his death-blow?" And the Ring that he held seemed to him exceedingly fair to look on; and he would not suffer it to be destroyed. (Tolkien 2007, 272)

Recalling our example of the *One Ring*, the *Ring*'s sole purpose is to dominate life, to take away choice and the possibility of moral victory, and once that device has been created, it becomes hard, if not impossible, to destroy. VSD aims to increase our abilities to be morally responsible, and it does so by increasing our choices, both now and in the future. In the next chapter, we will see some of the precise tools VSD provides for actually doing this in practice.

8

Ethics by Design in Action

> In designing tools we are designing ways of being – ways of being with moral and ethical import.
>
> Batya Friedman and David Hendry, "Value sensitive design: Shaping technology with moral imagination"

The Ethical Engineer's Toolbox

By this point we have seen in both real-world and fictional examples that technology is best understood as interactional. Moreover, the interactional nature of technology means that not just designers, but also stakeholders like ourselves, have a responsibility to participate in the design process and ensure that technologies align with our most fundamental values. The question we are now left with is how precisely to do that. How are stakeholders to be incorporated into design processes, and how are they, along with designers and engineers, to ensure that values shape and guide design?

In the previous chapter, we looked at various approaches used in technology design, highlighting some of the benefits and shortcomings of each, and concluding that the specific design needs of particular projects will often speak in favor of different approaches. That being said, we also saw that value sensitive design retains many of the benefits of the other approaches and has also incorporated a number of tools and methods which make it a more comprehensive design methodology (Friedman and Hendry 2019). In this chapter, we will examine in more detail the tools and methods of VSD, in order to see how one may actually engage in "ethics by design." This chapter goes into more detail on this very interesting and important topic, but if you don't need these details, you can skip ahead without loss of continuity. This chapter is best viewed as a sort of "toolbox" for applying the insights we have learned throughout this book, and the designer or engineer may return here for guidance on how best to implement values and ethics in design.

Who are the Stakeholders?

Value sensitive design is founded on the central awareness that real people are impacted on by technologies, and therefore should be given a voice in the design process. It is only in this way that designers can truly ensure they are aligning their decisions with the values of those being so impacted on. However, in order for designers to do this, they first need to know who the relevant stakeholders are, and VSD provides tools to do just that. Unlike Universal Design, VSD seeks not to design for everyone, but rather for those particular stakeholders who are impacted on by a system. VSD also pays heed to the severity of impact which a technology may have on different stakeholder groups. For this reason, it distinguishes between two classes of stakeholders. The first are *direct stakeholders,* which are those people or groups which directly interact with a system and are thus directly influenced by that system's effects. This can include users as well as designers themselves. The second group, (2) *Indirect stakeholders*, are those people and groups who do not directly interact with a technology but are nonetheless impacted on by it. This latter group is often not considered in other design approaches, and can include people from larger society, other firms or organizations which may be impacted on by the technology, or even nonhuman animals and the environment (Friedman et al. 2013).

Despite this generally more fine-grained way of categorizing those impacted on by technologies, we still need concrete ways of determining who specifically belongs to these groups so that they may effectively be brought into the design space. *Stakeholder Analysis* and *Stakeholder Tokens* present two ways designers can (more) straightforwardly identify who the stakeholders are for a particular system and determine how they interact with the system in question (Friedman et al. 2006; Yoo 2021). Let us first examine Stakeholder Analysis. The goal of Stakeholder Analysis is to ensure that the design process accounts for the values and needs of all stakeholders, rather than just those of designers or intended users (Watkins et al. 2013). This helps to create designs that consider the wider impact a technology will have on various people and groups. To conduct a Stakeholder Analysis, designers follow a few clear steps:[22]

1. *Identify all stakeholders*: Designers must first identify all individuals and groups who have a stake in the design process, including designers, users, funders, regulatory bodies, and any other groups that may be impacted on by design decisions.
2. *Assess the interests and values of each stakeholder*: After identifying stakeholders, designers must learn the values, goals, and motivations of them, as well as the potential impact designs might have on stakeholder interests.
3. *Prioritize stakeholders*: Once interests and values have been assessed, designers can prioritize them

based on the potential impact a design might have on those interests. This helps designers to focus on the most important stakeholders (i.e., those most impacted on by a design) and ensure that their needs are adequately addressed.

4. *Develop strategies for addressing the interests and values of stakeholders*: Based on the priorities identified in the previous step, designers can develop strategies for meeting the interests and values of each stakeholder. This may involve incorporating their input into the design process, negotiating with them to find mutually beneficial solutions, or finding ways to mitigate any negative impacts a design may have on their interests.

Stakeholder tokens, on the other hand, present a way for designers to identify and represent the values and interests of stakeholders. Stakeholder tokens can be physical or virtual objects that are used to symbolize the perspectives and concerns of different stakeholders, and can be used to facilitate discussions and decision-making around design decisions that impact on those stakeholders (Yoo 2021). To use stakeholder tokens, designers typically follow these steps:

1. Like *stakeholder analysis*, designers first *identify all stakeholders* (see above).
2. *Assign stakeholder tokens*: Each stakeholder is then assigned a physical or virtual token representing their interests, preferences, and values. These tokens can be as simple as plastic blocks or paper chips, or may be more complex, such as virtual avatars.
3. *Facilitate discussions and decision-making*: During the design process, designers can use stakeholder tokens to help facilitate discussion and decision-making around design decisions that impact on different stakeholders. For example, stakeholders can

use their tokens to indicate support or opposition to a particular design solution.

4. *Track stakeholder engagement*: By using stakeholder tokens, designers can track the level of engagement and input from different stakeholders during the design process. This can help designers to identify stakeholders who may not be adequately represented, and take steps to ensure that their interests and values are considered.

What are the Values?

Above, we said that once we know who the stakeholders are, designers must then grapple with those stakeholders' values, goals, and motivations, using these to develop strategies and designs which are sensitive to those values. However, how exactly are designers to determine what values some stakeholder (group) has? Sometimes, simply asking them can be as good a way as any, but designers may often be confronted with confusing or contradictory answers, complicating responsible design efforts. There are, however, ways to help designers elicit useful responses to these questions, and which further allow them to represent stakeholder values in a manner that is tailored to design. Though value sensitive design comes with a number of tools for doing this, we will focus on two which have proven to be popular among practitioners, namely *value scenarios* and *value-oriented semi-structured interviews*, both of which provide designers with tested methods for eliciting and representing the values of stakeholder groups (Czeskis et al. 2010).

The specific goal of value scenarios is to understand the potential impacts and trade-offs of different design decisions on the values of different stakeholders, and to identify design solutions that align with their values

(Nathan et al. 2008). To use value scenarios, designers typically follow these steps:

1. *Identify key values*: For each stakeholder, designers identify the values most important to them. These may include things like autonomy, privacy, security, accessibility, efficiency, or environmental sustainability.
2. *Develop value scenarios*: Based on the identified values, designers develop hypothetical scenarios that describe different design decisions and the potential consequences those would have for different stakeholders. For example, a scenario might describe a design decision that maximizes efficiency but sacrifices privacy, or a decision that prioritizes accessibility but reduces security.
3. *Evaluate value scenarios*: Once value scenarios have been developed, designers evaluate them to understand the potential impacts and trade-offs of different design decisions on the values of different stakeholders. This can help designers to identify design solutions that align with the values of all stakeholders.

Value scenarios place more emphasis on the efforts of designers to understand and incorporate stakeholder values into the design processes. *Value-oriented semi-structured interviews* (VOSIs), on the other hand, focus more on the stakeholders themselves, and involve conducting interviews with diverse groups including users, designers, and other relevant parties (Friedman 1997). These interviews are semi-structured, meaning that the interviewer has a set of guiding questions but may also follow up on topics and ideas that emerge during the conversation. The focus of such interviews is to understand the values and ethical considerations important to stakeholders, as well as how stakeholders might be affected by the product or service being designed. VOSIs can be an invaluable tool because they allow designers to gather a rich and diverse set of perspectives, providing the design process with a more complete and inclusive picture of stakeholder concerns. There are several steps that designers can follow when conducting value-oriented semi-structured interviews:

1. *Identify and select stakeholders*: The first step in conducting VOSIs is to identify which stakeholders should be included in the interviews. Users, designers, experts, and other relevant parties may all be included, but it is most important that a diverse group is chosen in order to allow for a wide range of perspectives, values, and ethical considerations.
2. *Develop a list of guiding questions*: Once stakeholders have been identified, designers must develop a list of guiding questions to structure the interviews. These questions should be formulated to help the designer understand the values and ethical considerations that are important to stakeholders, as well as how they might be affected by the product or service being designed.
3. *Conduct the interviews*: VOSIs are typically conducted in person or over the phone and can be recorded for later transcription and analysis. It is important to create a comfortable and relaxed atmosphere for the interview, as this will encourage the stakeholder to share their thoughts and opinions more openly.
4. *Analyze the data*: After interviews have been conducted, the collected data must be analyzed. This might involve transcribing the interviews and coding the data according to themes or categories. The data can then be further analyzed to identify common patterns and trends in the values and ethical considerations that were discussed.

5. *Use the data to inform design decisions*: The insights and perspectives gathered through VOSIs can be used to inform the design process and guide the development of products and services that are more aligned with the values and ethical considerations of stakeholders.

How Do we Use Values?

Once they know the stakeholders' values, designers must determine which ones are to be incorporated into design and, more importantly, how to go about translating those values into design requirements that can actually be designed for. As with the previous steps, there are several tools to aid designers in this process. Let us focus on two in particular, *value dams and flows* and *value hierarchies*.

Value dams and flows is founded on the idea that values can be thought of as resources which can be protected (dammed) or expended (flowed). Such damming and flowing can be pursued at many points throughout the design process, and the mentality which goes with value dams and flows can help designers understand the values at stake and how they may be affected by the product or service being designed (Denning et al. 2010). To use this method, designers first identify the values relevant to the design process, which might include things like privacy, security, or accessibility. Next, designers map out the various stages of the design process and consider how each stage may affect the values that have been identified. For example, a designer might consider how a new feature might impact on the privacy of users, or how a new material choice might affect the sustainability of the product. More specifically, to use this method designers can follow the steps below:

1. *Identify relevant values*: First, designers must identify the values relevant to the design process, which may include things like privacy, security, accessibility, or sustainability.
2. *Map out the design process*: Next, designers map out the stages of the design process, including key decisions and actions that will be taken at each stage.
3. *Identify value dams and flows*: For each stage of the design process, designers consider how the values identified (in step 1) may be affected by the decisions and actions they are taking. This might involve identifying value dams (ways in which the values are protected) or value flows (ways in which the values are expended).
4. *Evaluate the value dams and flows*: Designers evaluate the value dams and flows to determine which values are being protected or expended, and to what extent. This can help them make more informed and ethically sound design decisions.
5. *Use the value dams and flows to inform design decisions*: The insights gained from the value dams and flows analysis can be used to inform the design process and guide the development of products and services that are more aligned with stakeholders' values.

Where value dams and flows show how value may be fostered or squandered in an absolute or ordinal way, value hierarchies, on the other hand, are used to represent the relative importance or priority of different values in a design process (i.e., a cardinal viewing of values). A value hierarchy can help designers understand how different values relate to one another (in terms of rank) for different stakeholders, which in turn can serve as a guide for decision-making in the design process (Van de Poel 2013; Umbrello and Van de Poel

2021). In particular, value hierarchies aim to help designers turn values, which are often abstract, into more concrete technical design requirements. To create a value hierarchy, designers first identify the values that are relevant to the design process. Next, they work with stakeholders to determine the relative importance or priority of each value in relation to others. This can be done through methods like value-oriented semi-structured Interviews, focus groups, or surveys. Once the relative importance of values has been determined, designers can create a values hierarchy by ranking the values in order of importance. The values at the top of the hierarchy are considered the most important, while those at the bottom are considered the least important. To utilize a values hierarchy, a designer can follow these steps:

1. *Create a values hierarchy*: Once the relative importance of values has been determined, designers create a value hierarchy by ranking the values in order of importance. The values at the top of the hierarchy are considered most important, and those at the bottom are considered least important.
2. *Translate values into design requirements*: Using the value hierarchy as a guide, designers can translate values into specific design requirements. For example, if privacy is a high priority value, designers may specify that the product or service must have robust privacy controls. Similarly, if accessibility is a high priority value, designers might specify that the product or service must be easily accessible for users with disabilities.
3. *Evaluate and revise the design*: As the design process progresses, designers can continue to use the value hierarchy to evaluate designs and make revisions to ensure that the design remains aligned with stakeholders' values.

Stakeholders, Time, Values, and the Pervasiveness of Technology

The tools so far mentioned can be invaluable during what could be called a value sensitive design cycle. Each is suited to a unique and specific part of a design program that places stakeholder values at its center. Of course, such a cycle will never be perfect (and as we will see in the final chapter, that should not even be our goal), as many of these tools will need time, training, and sometimes specialized expertise in order to be employed effectively. Thus, despite their usefulness, designers will sometimes need more efficient approaches that are less time- and resource-intensive. One such tool for meeting these dual challenges of efficiency and faithfulness to values is *envisioning cards* (Friedman and Hendry 2012; Yoo et al. 2022; Umbrello 2022c).

Envisioning cards are used in value sensitive design to help designers in exploring and considering the values and (ethical) implications of a product or service (Friedman and Hendry 2012). They are a set of thirty-two cards, each with a different value or ethical consideration written on it, that can be used in a variety of ways to stimulate discussion and reflection among designers and stakeholders. The cards are categorized into four different suites:

1. *Stakeholder cards* help designers determine who the *direct* and *indirect stakeholders* could be and their different potential perspectives.
2. *Time cards* aid designers in thinking about how their work may affect the future. Until the technology has advanced past the initial stages of novelty and is fully assimilated into society, these ramifications might not become obvious. The cards aid in ensuring that technology will have a favorable and long-lasting

impact on society by instructing designers to consider the longer-term impacts of their ideas.

3. *Value cards* help designers consider how values are impacted on by a system, and vice versa. Designers can use these cards to stimulate discussion around and reflection on the values that should be considered in the design process.
4. *Pervasiveness cards* help designers focus on what widespread adoption and use of a technology might mean for society, and how society might then impact on the technology. Different factors, such as geography, culture, and demographics, might contribute to the spread of technologies. The Pervasiveness criterion aids designers in taking into account the broader implications and effects of their technology on various individuals and societies.

The main set also comes with a supplementary set of cards with a fifth suite: *multi-lifespan cards*. These cards support designers of tools, technology, policy, and infrastructure to think about and address serious societal issues that might not have quick fixes (Yoo et al. 2022). The goal of this criterion is to assist designers in addressing issues that call for a more comprehensive, multi-lifespan approach to design and that could have long-term effects. The multi-lifespan criterion can assist designers in producing solutions that are more sustainable and have a good influence on society by encouraging consideration of these complicated challenges. Policymakers in particular can use the cards to help co-create technology policy. For example, climate change is a complex, multi-faceted issue that will affect multiple generations. Quick "fixes," such as minor reductions in greenhouse gas emissions or employment of climate engineering technologies risk potentially dangerous long-term consequences.

In this context, a policymaker might use the multi-lifespan cards to think about the long-term impacts of different policy decisions. For example, they might use the "weaving social fabric" card to consider how today's decisions will affect people living several generations from now and their complex social milieu. They might use the "collect now" card to think about what kind of world we want to leave for future generations and leave them a record of solution and knowledge that we found in the present. And they might use the "pause" card to consider how inaction, rather than action, is sometimes the best course of action when talking about intervening in complex systems like the global climate.

Some common ways to use envisioning cards in VSD include:

1. *Value brainstorming*: Designers can use envisioning cards to generate ideas and stimulate discussion around the values that should be considered during the design process.
2. *Value prioritization*: Designers can use envisioning cards to rank the relative importance of different values, helping to inform design decisions and prioritize design requirements.
3. *Value scenario planning*: Designers can use envisioning cards to explore different scenarios and consider the potential values and ethical implications of each scenario.
4. *Value trade-off analysis*: Designers can use envisioning cards to consider the trade-offs that may be necessary in order to align the design with different values.

Tools like envisioning cards encourage designers to consider explicitly the values and ethical implications of their designs, which can lead to more ethically sound and value-based decision-making (McMillan 2019). Beyond this, they naturally stimulate discussion and collaboration among designers and stakeholders, helping to ensure that a wide range of perspectives is taken into account in the design process. This, in turn, can encourage designers to be imaginative and mindful of various possibilities and trade-offs, which can result in more creative new designs (Logler et al. 2018). Similarly, by clearly taking values and ethical issues into account, envisioning cards can help encourage transparency in the design process and make sure that stakeholders are conscious of the values that are being emphasized (Dexe et al. 2020).

Always Coming Back

The main aim of this chapter was to provide a sort of "field manual" or "toolbox" for designers and engineers who wish to do ethics-by-design. Naturally, no single chapter, nor book for that matter, can be entirely comprehensive in terms of all the tools, applications, and benefits that any given approach brings. However, this chapter presents a helpful starting point which practitioners can continually come back to when embarking on projects which aim to incorporate ethics into design.

9

Our Common Future with Technology

We began this book by examining three classic conceptions of technology, namely *instrumentalism*, *determinism*, and *constructivism*. Each provided some fundamental insights but, one by one, each also showed itself to be a source of error. However, by recognizing the advantages of these conceptions, we were able to locate the building blocks of *interactionalism*, which is the foundation of the contemporary ethics of technology and design.

Instrumentalism sees technologies as mere tools or means for achieving specific ends or goals. It is a perspective which emphasizes the practical and functional aspects of technology and focuses on how technology can be used to solve problems, improve efficiency, and achieve specific objectives. One of the key things this view gets right is its recognition of the practical and functional value of technology; technology often is developed and used for specific purposes, and it can be an effective and powerful tool for achieving a wide range of goals and objectives. Still, instrumentalism has its limitations. One of its main criticisms is that it tends to oversimplify the complex and dynamic relationships between technology and society. By focusing solely on the practical and functional aspects of technology, one loses sight of the many other ways in which technology shapes and is shaped by social, cultural, and political factors. Moreover, instrumentalism often ignores the fact that technology is not neutral or value-free. Both the development and use of technologies are always shaped by the values and priorities of the people and institutions involved, and technology can have both positive and negative consequences for society. By failing to consider the social and cultural implications of technology, the instrumental view can thus contribute to narrow and oversimplified understandings of the role of technology in society.

The second conception, technological determinism, is the view that technology and technological developments are the primary driving forces behind social, economic, and cultural change. This perspective maintains that technologies are the main factors shaping individual human behavior and even society as a whole. In a sense, there is truth to this notion, as technologies can have a powerful and far-reaching influence on society. New technologies often profoundly impact on the way people live, work, and interact with each other, and can lead to significant changes in social norms and cultural practices. The industrial revolution, for example, was driven in large part by technological innovations such as the steam engine and the power loom, inventions which led to significant changes in the way that goods were produced and consumed, and these economic changes further led to sweeping shifts in the organization of societies and the ways people lived. Similarly, the development of computers and other digital technologies has enabled the growth of the information economy and has created new opportunities for innovation and economic growth. However, there are significant downsides to viewing technology through this deterministic lens.

One of the main limitations of this perspective is its tendency to oversimplify the complex and dynamic relationship between technology and society. By suggesting that technology is the primary driving force behind social change, determinism ignores the many other factors which contribute to social and cultural change, including political, economic, and cultural forces. Additionally, technological determinism often ignores the fact that people have agency and are not simply passive recipients of technological change. People have the ability to shape and influence the way that technology is developed and used, and to make choices about the role that technology plays in their lives.

The social construction of technology (or social constructivism) rose as a response to determinism, and sees technology as being shaped and influenced by social, cultural, and political factors. This perspective emphasizes the idea that technology is not a neutral or objective phenomenon, but rather a product of the social and cultural contexts in which it is developed and used. One of the central insights of social constructivism is the recognition that technology is not a fixed or static phenomenon, but rather a constantly evolving and changing process shaped by the people and institutions involved in its development and use. However, social constructivism also has its share of shortcomings. One of the main critiques is that it tends to minimize the role of technical and scientific factors in shaping the development and use of technology, suggesting that technology is primarily a social and cultural construct, when in fact the technical and scientific aspects can be just as important and influential. Furthermore, constructivism often ignores the fact that technology can have powerful and far-reaching impacts on society; while technology is shaped by social and cultural factors, it also shapes and influences those factors in turn. This latter insight led us to explore an amalgamation of these three approaches, leading us to an *interactional* conception of technology.

Technology is not just a tool or means to achieving a specific end, but rather a complex and dynamic system that shapes and is shaped by people and societies. The ways people use technology and the ways technology shapes their behavior and choices are constantly evolving processes influenced by a wide range of factors. It is in this sense that technology is best understood as interactional, as technology and society do not exist apart, but persistently impact on one another. Interactionalism further highlights how many different factors influence technology, emphasizing the importance of understanding the broader context in which technology is developed and used. The interactional perspective also pays heed to the role of human agency in shaping technological change, recognizing that people are not simply passive recipients of this, but have the ability to shape and influence the development and use of technologies. This offers a valuable and nuanced perspective on the relationship between technology and society which is more comprehensive and multi-dimensional than alternative views of technology can provide.

Returning to our running example of Tolkien's *One Ring*, the central ideas of interactionalism can be seen in many aspects of this fictional device. The *Ring* exerts a powerful and often dangerous influence over those who wield it, manipulating their thoughts and actions and often leading them to make decisions they would not normally make or behave contrary to their nature. This presents a metaphor for technology's sometimes controlling or manipulative influence over people. On the other hand, the *One Ring* is a highly coveted and sought after object for beings both good and evil. This desire to possess both it and the power and control it represents can be seen as a reflection of the way people often desire and seek out new technologies, even those which might have negative consequences or risks associated with them. And finally, the *One Ring* is an object shaped by its maker, a technological device imbued with the values and intent of the one who crafted it. Tolkien's *One Ring* can thus be seen as a useful illustration of the complex and dynamic relationship which exists between people and technology and which is at the core of the interactional perspective.

Technology ethics is about taking this interactional perspective seriously and using it as a foundation for responsibly innovating. To do this in practice, the interactional stance relies on, among other things, the method of *value sensitive design*, an approach to designing technology that takes into account the preferences, priorities, and moral values of the people affected by a technology. This approach emphasizes that technology impacts heavily on individuals' lives, as well as on what future technologies will emerge (which also impact on individuals' lives, and on future technologies…), and as such, technologies should be designed to be sensitive to the core values of not just the people who will be using it, but also any other affected parties. In paying heed to the broader context of a technology's development and use, as well as the values embodied in technology, interactionalism is closely aligned with this approach to design. Beyond this, interactionalism's emphasis on the role of human agency in shaping technological development and use leads naturally to value sensitive design, which focuses on considering the needs and preferences of stakeholders and on bringing them into the design process wherever possible. Thus, while they are not the same, interactionalism and value sensitive design bear close connection, and both point toward a richer and more optimistic view of both technology and our role in its future.

Technological Ethics in Practice

The optimistic outlook fostered by interactionalism and value sensitive design is tempered by practical considerations encompassing all facets of technology ethics. These considerations shape our approach to the ethical challenges posed by technological advancement and guide us in implementing the frameworks discussed earlier.

Practical challenges emerge from the need to balance diverse stakeholder interests, which include developers, users, and society at large. Technology designers and policymakers must weigh these interests against the social, cultural, and environmental impacts of technological innovation, considering how these impacts might evolve. Staying informed about the latest developments in technology ethics is crucial for all involved in the creation and governance of new technologies. Conflicts of interest are a universal concern in the field of technology. Financial incentives or other pressures can influence decisions, potentially leading to outcomes that do not align with societal or ethical standards. Recognizing these conflicts and planning for them from the outset is critical to ensuring that technologies are developed responsibly.

Addressing these challenges requires concerted efforts at both institutional and individual levels. Codes of conduct and ethical guidelines can orient the design process toward ethical considerations. Independent ethics committees or advisors can provide support for navigating ethical dilemmas before they escalate. Incorporating ethics education into technology-related curricula ensures that future professionals are equipped to tackle these issues. Mechanisms for enforcing ethical standards, such as certification requirements and professional review boards, can further ensure adherence to ethical practices. Engaging in dialogue with stakeholders and fostering collaborative environments allows for a multiplicity of perspectives to be considered in the decision-making process.

Technology ethics is integral to maintaining public trust; breaches of ethical conduct can erode confidence in technological professions and innovations. Incidents like the Volkswagen emissions scandal (Hotten, 2015) demonstrate how ethical lapses can have far-reaching consequences, emphasizing the real-world significance of maintaining rigorous ethical standards in technology development. The value of interactionalism becomes evident when considering these ethical constraints. It recognizes that sociotechnical systems are inherently complex, and that managing the interplay between technology, society, and the environment requires a flexible, responsive approach. As we navigate the evolving landscape of technology, it is essential to continuously adapt our ethical frameworks to ensure that they remain relevant and effective in promoting the common good.

Progress, not Perfection

As a closing thought for this book, it is important that we accept that we will not always be able to create perfect technologies which perfectly align with all our values. And, even supposing we could, our values change, so sooner or later, our technologies would no longer reflect them. Perfection simply cannot be achieved, and in fact, *it should not be the goal of design*. Instead, we should aim for *progress*.

The ethos of "progress, not perfection" finds its analogy in the construction of the Duomo di Firenze, or Florence Cathedral. The Duomo is a famous and iconic example of Gothic architecture, but its construction spanned hundreds of years, with building beginning in the mid thirteenth century and not being completed until the late fifteenth century. During that time, the cathedral's builders and designers employed an approach best described as one of "progress, not perfection"; rather than aiming to complete the entire construction all at once or do so in a way that was perfectly true to the plans, the builders of the Duomo focused on making steady and consistent progress over whatever period of time was required. This allowed the building to evolve and change as the needs and priorities of the community changed, resulting in a structure that is both beautiful and enduring.

In the world of technology, the approach of continuous iteration and feedback resembles the "progress, not perfection" ideology. Modern software development, for example, often employs Agile methodologies. Agile development is characterized by short "sprints" or phases where feedback is constantly sought and incorporated, ensuring the software improves over time. This iterative process, much like the construction of the Duomo, emphasizes progress and evolution over rigid, one-time perfection. Just as the builders of the Duomo trusted future generations to refine and complete their vision, engineers today should embrace the dynamic nature of design, recognizing that products will need to evolve in response to changing societal needs and values. Moreover, trust once lost can have cascading effects. After incidents like the Cambridge Analytica scandal, where personal data of millions was harvested without consent, public trust in digital platforms like Facebook took a significant hit (Confessore 2018). This wasn't just a challenge for the company but for engineers and innovators in the broader tech industry. Such instances reiterate the importance of ensuring ethics is at the forefront of all engineering endeavors.

In our pursuit of progress, we must acknowledge, as Tolkien insightfully notes, that "Even the very wise cannot see all ends" (Tolkien 2004, 59). This humbling reminder cautions us against the presumption that we can predict all the outcomes of our technological endeavors or that our current understanding of "good" will remain unchallenged by time and experience. Whenever we evaluate technological evolution, it is essential to recognize "progress" as a guiding ethos rather than a destination. The notion of progress in technology ethics is not merely about technological advancements or milestones achieved but about the continual alignment of our creations with the shifting landscape of societal perception of values. This concept of progress is inherently fluid, adapting to the ebb and flow of human experience and understanding. It acknowledges that what constitutes progress today may need reevaluation tomorrow. By instilling a mindset that welcomes change and values adaptability, we lay the groundwork for technologies that not only serve our present needs but are also malleable to the moral progress and cultural shifts of future societies. In this way, progress becomes a perpetual cycle of reflection, action, and refinement — a commitment to ethical stewardship that transcends the limitations of our present capabilities and aspirations.

Aiming for progress rather than perfection is also closely connected to the interactional view; since technologies are complex and dynamic systems shaped by their interactions with people and the larger world, it is never possible to arrive at eternally satisfactory solutions. Instead, like the Duomo, technologies must evolve and change as the needs of those for whom they are designed evolve and change. The idea of "progress, not perfection" is thus an important reminder that technology cannot be completed or perfected in a single moment, but is an ongoing process requiring ongoing effort and attention. The architects and patrons of the Duomo had a dream which they knew they would never see realized, and in that knowledge, they handed their dream down to future generations, trusting that those who followed would see it through. They made a leap of faith, but it is a leap that all ethically oriented engineers will have to take at some point. The simple fact is that technologies do not become "done" simply because we are done with them. They are put into the world, and they shape the world, and we must be vigilant in our efforts to continually ensure that they are shaping it for the better.

Glossary

Artificial Intelligence (AI)

A class of technologies that are autonomous, interactive, adaptive, and capable of carrying out human-like tasks, in particular, as a function of Machine Learning, which allows such technologies to learn on the basis of interaction with (and feedback from) the environment.

Direct Stakeholders

An individual or group who interacts directly with a technology. For example, a system of electronic medical records might be designed for doctors and insurance companies. See stakeholder, indirect.

Ethics by Design

An approach to technology ethics and a key component of responsible innovation that aims to integrate ethics in the design and development stage of the technology. Sometimes formulated as "designing for values" or "values at play." Similar terms are "value sensitive design" and "ethically aligned design." For use of the latter phrase, see the IEEE project on ethically aligned design for AI / ML systems (Shahriari and Shahriari 2017).

Framing Effects

A cognitive bias in which individuals select items based on whether they have positive or negative implications.

Human-Centered Design

A design approach focusing on the requirements, preferences, and skills of the users of a good, service, or system.

Indirect Stakeholders

An individual or group who is impacted on by a technology but does not directly interact with it.

Neutrality Thesis

The neutrality thesis concerning technologies holds that technologies are not the fundamental bearers of values but rather that technologies are only valuable in an instrumental way.

Participatory Design

An approach to design that aims to involve in the design process as many stakeholders as possible who are impacted on by a design. This is done in order to ensure that the resulting system is aligned with and meets their needs.

Universal Design

An approach to design that aims to make the product accessible to all groups regardless of ability, age, or other contingent factors.

Value Sensitive Design

A principled approach to "ethics by design," often abbreviated as VSD, employing co-design tools and methods to include the various values of both direct and indirect stakeholders early on and throughout the design and deployment process of a system's design.

Further Reading

Asveld, Lotte, Rietje van Dam-Mieras, Tsjalling Swierstra, Saskia Lavrijssen, Kees Linse, and Jeroen van den Hoven (2017) *Responsible Innovation 3: A European Agenda?* Cham: Springer.

Blok, Vincent (2023) *Putting Responsible Research and Innovation into Practice: A Multi-Stakeholder Approach*. Cham: Springer.

Borgmann, Albert (1984) *Technology and the Character of Contemporary Life: A Philosophical Inquiry*. Chicago: Chicago University Press.

Coeckelbergh, Mark (2020) *The Political Philosophy of AI*. Cambridge, UK: Polity.

Friedman, Batya, and David G. Hendry (2019) *Value Sensitive Design: Shaping Technology with Moral Imagination*. Cambridge, MA: MIT Press.

Gianni, Robert, John Pearson, and Bernard Reber (2019) *Responsible Research and Innovation: From Concepts to Practices*. 1st edn. London: Routledge.

Hare, Stephanie (2022) *Technology Is Not Neutral: A Short Guide to Technology Ethics*. London: London Publishing Partnership.

Jasanoff, Sheila (2016) *The Ethics of Invention: Technology and the Human Future*. New York: W. W. Norton & Company.

Kim, JaHun, Elaine Walsh, Kenneth Pike, and Elaine A. Thompson (2020) "Cyberbullying and victimization and youth suicide risk: The buffering effects of school connectedness." *Journal of School Nursing* 36 (4): 251–257.

Koops, Bert-Jaap, Ilse Oosterlaken, Henny Romijn, Tsjalling Swierstra, and Jeroen van den Hoven (2015) *Responsible Innovation 2: Concepts, Approaches, and Applications*. Cham: Springer.

Lonergan, Bernard (1992) *Insight: A Study of Human Understanding, vol. 3, Collected Works of Bernard Lonergan*. Toronto: University of Toronto Press.

Mitcham, Carl (2020) *Steps Toward a Philosophy of Engineering: Historico-Philosophical and Critical Essays*. London: Roman and Littlefield.

Nyholm, Sven (2023) *This Is Technology Ethics: An Introduction*. Hoboken, NJ: Wiley-Blackwell.

Peterson, Martin (2017) *The Ethics of Technology: A Geometric Analysis of Five Moral Principles*. New York: Oxford University Press.

Pinch, Trevor J., and Wiebe E. Bijker (1984) "The social construction of facts and artefacts: Or how the sociology of science and the sociology of technology might benefit each other." *Social Studies of Science* 14 (3): 399–441.

Robison, Wade L. (2016) *Ethics Within Engineering: An Introduction*. London: Bloomsbury Academic.

Rocco, Roberto, Amy Thomas, and María Novas-Ferradás (2022) *Teaching Design for Values: Concepts, Tools and Practices*. Delft: TU Delft Open.

Sandler, Ronald L. (2014) *Ethics and Emerging Technologies*. London: Palgrave Macmillan.

Spiekermann, Sarah (2023) *Value-based Engineering – A Guide to Building Ethical Technology for Humanity*. Berlin: De Gruyter.

Steen, Marc (2012) *Ethics for People Who Work in Tech*. London: Chapman & Hall.

Tromp, Nynke (2019) *Designing for Society: Products and Services for a Better World*. London: Bloomsbury Visual Arts.

Vallor, Shannon (2016) *Technology and the Virtues: A Philosophical Guide to a Future Worth Wanting*. Oxford: Oxford University Press.

Van de Poel, Ibo, and Lamber Royakkers (2011) *Ethics, Technology, and Engineering: An Introduction*. Hoboken, NJ: Wiley-Blackwell.

Van den Hoven, Jeroen, Neelke Doorn, Tsjalling Swierstra, Bert-Jaap Koops, and Henny Romijn (2014) *Responsible Innovation 1: Innovative Solutions for Global Issues*. Dordrecht: Springer.

Van den Hoven, Jeroen, Seumas Miller, and Thomas Pogge, eds. (2017) *Designing in Ethics*. Cambridge: Cambridge University Press.

Verbeek, Peter-Paul (2005) *What Things Do: Philosophical Reflections on Technology, Agency, and Design*. University Park, PA: Pennsylvania State University Press.

Winner, Langdon (2020) *The Whale and the Reactor: A Search for Limits in an Age of High Technology*, 2nd edn. Chicago: University of Chicago Press.

Zhang, Jiehuang, Ying Shu, and Han Yu (2022) "Fairness in design: A framework for facilitating ethical artificial intelligence designs." *International Journal of Crowd Science* 7 (1): 32–39.

Notes

1 The three chapters (2–4) reviewing instrumentalism, determinism, and social constructivism will issue in this position (Chapter 5) that then serves as the foundation for the practice-oriented chapters (6–8) that follow.

2 Instrumentalism, determinism, and constructivism are terms used across various other philosophical disciplines. In what follows, for the sake of brevity, we will use "instrumentalism," "determinism," and "constructivism" as specifically applied to technology.

3 See also Latour, *Pandora's Hope* (1999), 176 and Verbeek, *What Things Do* (2005), 155 which are cited therein.

4 Note on the use of Tolkien analogies: The references to J.R.R. Tolkien's *The Lord of the Rings* throughout this manuscript are not solely for their specific cultural relevance, but rather for the timeless and universal themes they embody. Tolkien's work delves into moral dilemmas, the interplay of individual choices with larger systems, and the ever-present struggle between opposing forces – themes that resonate deeply with the ethical challenges explored in this book. For readers unfamiliar with Tolkien's world, the objective here is not a detailed exposition of his narratives but a drawing from the broad strokes of his themes as a lens through which we can better understand the complexities of technology ethics. Whether or not one is acquainted with middle-earth, the central messages of these analogies aim to be both accessible and illuminating.

5 Even the UN has adopted this position, viewing technology neither as purely deterministic nor instrumental, and instead affirming the interactional nature of technology and social

factors, thereby permitting a holistic approach to finding solutions, see Umbrello (2023a),

6 Unlike methane (produced by livestock), which remains in the atmosphere for a relatively short period of time, in vitro meat produces large quantities of greenhouse gases in the form of carbon dioxide, which remains for hundreds of years, Carni Sostenibili (2021).

7 A tweet composed and published by David Gunkel on Twitter on January 18, 2023 concerning the burgeoning deterministic discourse on OpenAI's Chat GPT program.

8 Compare with Brunner (1887).

9 For a dry, yet comprehensive overview of the method, the reader is directed to Pinch and Bijker (1984) in Further Reading.

10 More controversially, nudges may also be used to foster certain behaviors which are not necessarily good for the actor being nudged, but rather which are deemed economically or societally useful. For work highlighting this potential worry, see Lavi (2018) and Glod (2015).

11 Tolkien, *The Silmarillion*, 265.

12 Abraham Lincoln, quoted in Coleman (2022). Note that though this quote has been attributed to Abraham Lincoln, there is no concrete evidence he said this, but the Abraham Lincoln Presidential Library and Museum states that "the sentiment clearly matches other things President Lincoln is documented to have said" (see McKinney 2018).

13 In particular, the assessments of the Human Genome Project provided much impetus for the movement toward responsible innovation. See (Felt, 2018).

14 In fact, RRI was a key component of "Horizon 2020," the European Commission's agenda for funding science (Pain 2017).

15 This, we may recall from Chapter 2, is the *Collingridge Dilemma*, that in order to see the problems associated with technologies, it is often necessary to first see those technologies in society, "in action," as it were, but by that point, the possibility for altering or fixing technologies may be greatly limited by the ways those technologies have impacted on society, for better or worse.

16 According to Google's own statistics, in 2016 only 3% of all Android smartphones ran the most recent operating system, and only half of those devices received a security patch, see Greenberg (2017).

17 For an example along these lines, see (Manancourt 2020) for discussion of Norway's scuttling of its digital contact ad tracing platform in favor of a policy-based solution instead.
18 We can already recall stories from antiquity cautioning us about both the impacts of technology on society and technological overreach. Examples include the myth of Icarus from Greek mythology, the golem from Jewish folklore, and the Tower of Babel from the Bible's Book of Genesis.
19 The term "human-centered design" was first coined by Don Norman in his 1988 book *The Design of Everyday Things*.
20 Compare with Umbrello (2022c).
21 The objectivity of moral values in value sensitive design is best understood as Bernard Lonergan understands it, that is, as involving an active and self-correcting process of inquiry that takes into account empirical evidence, rational analysis, personal experience, and the perspectives of others, that is "the fruit of an authentic subjectivity," see Egan (2009, 166). For a more in-depth discussion of this, see Lonergan (1992, 170). I have already begun delineating how this can be done for value sensitive design, see Umbrello (2023c).
22 Some accounts of VSD may present slightly different methods or give different weightings or orderings to the steps. The presentation here should thus not be taken to be exhaustive or necessarily comprehensive, but rather as merely an indicator of the general shape of *stakeholder analysis* as it is found across VSD accounts. For more specific exploration of this tool, see, e.g., Friedman and Hendry (2019, ch. 3).

References

Adomaitis, Laurynas, Alexei Grinbaum, and Dominic Lenzi (2022) "TechEthos D2.2: Identification and specification of potential ethical issues and impacts and analysis of ethical issues of digital extended reality, neurotechnologies, and climate engineering." *TechEthos Project Deliverable*. Available at: www.techethos.eu

Albergotti, Reed (2019) "Apple accused of worker violations in Chinese factories." *Washington Post*. 9 September. https://www.washingtonpost.com/technology/2019/09/09/apple-accused-worker-violations-chinese-factories-by-labor-rights-group/

Alleman, Mark (2000) "The Japanese Firearm and Sword Possession Control Law: Translator's Introduction." *9 Pacific Rim Law & Policy Journal*: 165.

Asaro, Peter M. (2000) "Transforming society by transforming technology: the science and politics of participatory design." *Accounting, Management and Information Technologies* 10 (4): 257–290.

Bannon, Liam J., and Pelle Ehn (2012) "Design: design matters in Participatory Design." In *Routledge International Handbook of Participatory Design*, eds. Jesper Simonsen and Toni Robertson, 37–63. Milton Park, UK: Routledge.

BBC (2012) "NRA: 'Good guys with guns stop bad guys with guns'." *BBC News*, December 12. https://www.bbc.com/news/av/world-us-canada-20817967

BBC (2022) "Halewood man who killed burglar after seeing break-in via Ring doorbell jailed." *BBC News*, August 5. https://www.bbc.com/news/uk-england-merseyside-62439803

Benthall, Jonathan (1976) *The Body Electric: Patterns of Western Industrial Culture*. London: Thames & Hudson.

Beroggi, Giampiero E. G. (2008) "Secure and easy internet voting." *Computer* 41 (2): 52–56.

Bimber, Bruce (1990) "Karl Marx and the three faces of technological determinism." *Social Studies of Science* 20 (2): 333–351.

Bombas (n.d.) "Bombas giving." *Bombas*. https://bombas.com/pages/giving-back

Bose (n.d.) "Bose sleepbuds II." *Bose*. https://www.bose.com/en_us/better_with_bose/better-sleep-is-back-with-bose-sleepbuds.html

Bostrom, Nick (2005) "A history of transhumanist thought." *Journal of Evolution and Technology* 14 (1): 1–25.

Brunner, Heinrich (1887) "Der Reiterdienst und die Anfänge des Lehnwesens." Zeitschrift der Savigny-Stiftung für Rechtsgeschichte. *Germanistische Abteilung* 8 (1): 1–38.

Buchinger, Eva, Manuela Kinegger, Georg Zahradnik, Michael J. Bernstein, Andrea Porcari, Gustavo Gonzalez, Daniela Pimponi, and Guiliano Buceti (2022) "TechEthos technology portfolio: Assessment and final selection of economically and ethically high impact technologies." *Deliverable 1.2 to the European Commission*. TechEthos Project Deliverable. Available at: www.techethos.eu.

Carlsson, Sven (2020) "Tegnell säger nej till smitt-spårning via mobilen." *Sveriges Radio*, May 2. https://sverigesradio.se/artikel/7464370

Carni Sostenibili (2021) "Lab-grown meat is less sustainable than you think." *Carni Sostenibili.* https://www.carnisostenibili.it/en/lab-grown-meat-is-less-sustainable-than-you-think/#:~:text=While%20livestock%20emits%20methane%2C%20a,atmosphere%20for%20hundreds%20of%20years

Caro, Robert A. (1975) *The Power Broker: Robert Moses and the Fall of New York*. New York: Vintage Books.

CEUD (n.d.) "The 7 Principles." *Centre for Excellence in Universal Design.* https://universaldesign.ie/What-is-Universal-Design/The-7-Principles/?fbclid=IwAR3jlmv5BdPeY36eYKqd3u-wnRLPA_mNDju3CKrjbZf-MSlzavNqdJ-ulMg

Coeckelbergh, Mark (2017) *New Romantic Cyborgs: Romanticism, Information Technology, and the End of the Machine*. Cambridge, MA: MIT Press.

Coeckelbergh, Mark (2020) *AI Ethics*. Cambridge, MA: MIT Press.

Coleman, Nikki (2022) *Military Space Ethics*. Hampshire, UK: Howgate Publishing Limited.

Confessore, Nicholas (2018) "Cambridge Analytica and Facebook: The scandal and the fallout so far." *New York Times*, 4 April. https://www.nytimes.com/2018/04/04/us/politics/cambridge-analytica-scandal-fallout.html

Cumbo, Bronwyn, and Neil Selwyn (2022) "Using participatory design approaches in educational research." *International Journal of Research & Method in Education* 45 (1): 60–72.

Czeskis, Alexei, Ivayla Dermendjieva, Hussein Yapit, Alan Borning, Batya Friedman, Brian Gill, and Tadayoshi Kohno (2010) "Parenting from the pocket: Value tensions and technical directions for secure and private parent–teen mobile safety." *Proceedings of the Sixth Symposium on Usable Privacy and Security*: 1–15.

D'Amore, Bruno, and Silvia Sbaragli (2019) *La*

matematica e la sua storia, vol. III, Bari: Edizioni Dedalo, 2019.

D'Angour, Armand (2021) *How to Innovate: An Ancient Guide to Creative Thinking*. Princeton, NJ: Princeton University Press, 2021.

De la Bellacasa, Maria Puig (2012) "'Nothing comes without its world': Thinking with care." *The Sociological Review* 60 (2): 197–216.

De la Bellacasa, Maria Puig (2017) *Matters of Care: Speculative Ethics in more than Human Worlds*, vol. 41. Minneapolis, MN: University of Minnesota Press.

De la Cruz Paragas, Fernando, and Trisha T. C. Lin (2016) "Organizing and reframing technological determinism." *New Media & Society* 18 (8): 1528–1546.

Deng, Mina, Kim Wuyts, Riccardo Scandariato, Bart Preneel, and Wouter Joosen (2011) "A privacy threat analysis framework: Supporting the elicitation and fulfillment of privacy requirements." *Requirements Engineering* 16 (1): 332.

Denning, Tamara, Alan Borning, Batya Friedman, Brian T. Gill, Tadayoshi Kohno, and William H. Maisel (2010) "Patients, pacemakers, and implantable defibrillators: Human values and security for wireless implantable medical devices." *Proceedings of the SIGCHI Conference on Human Factors in Computing Systems*: 917–926.

Department for Digital, Culture, Media and Sport (2020) "National Data Strategy" [online]. https://www.gov.uk/government/publications/uk-national-data-strategy/national-data-strategy

Deus Ex: Human Revolution (2011) Microsoft Windows [Game]. Montréal: Eidos-Montréal.

Deus Ex Wiki (n.d.) "#CantKillProgress." *Deus Ex Wiki*. https://deusex.fandom.com/wiki/CantKillProgress

Dexe, Jacob, Ulrik Franke, Anneli Avatare Nöu, and

Alexander Rad (2020) "Towards increased transparency with value sensitive design." In *Artificial Intelligence in HCI. HCII 2020. Lecture Notes in Computer Science, vol. 12217*, eds. Helmut Degen and Lauren Reinerman-Jones, 3–15. Springer, Cham.

Egan, Philip A. (2009) *Philosophy and Catholic Theology: A Primer*. Collegeville, MN: Liturgical Press.

European Commission (2013) "Options for strengthening responsible research and innovation – Report of the expert group on the state of art in Europe on responsible research and innovation." *Publications Office*.

Evans-Pritchard, Blake (2013) "Aiming to reduce cleaning costs." *Works That Work* 1, winter. https://worksthatwork.com/1/urinal-fly

Fairphone (n.d.) "Our impact." *Fairphone*. https://www.fairphone.com/en/impact/

Farndon, John (2010) "*48 The Stirrup*." *The World's Greatest Idea: The Fifty Greatest Ideas that have Changed Humanity*. London: Icon Books.

Feenberg, Andrew (2001) "Looking backward, looking forward: Reflections on the 20th century." *Hitotsubashi Journal of Social Studies* 33: 135–142.

Felt, Ulrike (2018) "Responsible research and innovation." In *Routledge Handbook of Genomics, Health and Society* (2nd edn.), eds. Sahra Gibbon, Barbara Prainsack, Stephen Hilgartner, and Janelle Lamoreaux, 108–116. Milton Park: Routledge.

Flanagan, Mary, Daniel C. Howe, and Helen Nissenbaum (2008) "Embodying values in technology: Theory and practice." In *Information Technology and Moral Philosophy*, eds. Jeroen van den Hoven and John Weckert, 322–353. Cambridge, Cambridge University Press.

Fortier, John C., and Norman J. Ornstein (2002) "The absentee ballot and the secret ballot: Challenges for election reform." *University of Michigan Journal of Law Reform* 36: 483.

Friedman, Batya (1996) "Value-sensitive design." *Interactions* 3 (6): 16–23.

Friedman, Batya (1997) "Social judgments and technological innovation: Adolescents' understanding of property, privacy, and electronic information." *Computers in Human Behavior* 13 (3): 327–351.

Friedman, Batya, and David Hendry (2012) "The envisioning cards: A toolkit for catalyzing humanistic and technical imaginations." *Proceedings of the SIGCHI Conference on Human Factors in Computing Systems*: 1145–1148.

Friedman, Batya, and Hendry, David G. (2019) *Value Sensitive Design: Shaping Technology with Moral Imagination*. Cambridge, MA: MIT Press.

Friedman, Batya, and Khan, Peter H. (2003) "Human values, ethics, and design." In *The Human-Computer Interaction Handbook: Fundamentals, Evolving Technologies, and Emerging Applications*, eds. Julia A. Jack and Andrew Sears, 1177–1201. Mahwah, NJ: Lawrence Erlbaum.

Friedman, Batya, Peter H. Kahn Jr., Jennifer Hagman, Rachel L. Severson, and Brian Gill (2006) "The watcher and the watched: Social judgments about privacy in a public place." *Human–Computer Interaction* 21 (2): 235–272.

Friedman, Batya, Peter H. Kahn, Alan Borning, and Alina Huldtgren (2013) "Value sensitive design and information systems." In *Early Engagement and New Technologies: Opening Up the Laboratory*, eds. Neelke Doorn, Daan Schuurbiers, Ibo van de Poel, and Michael E. Gorman, 55–95. Springer, Dordrecht.

Gabberty, James W., and Robert G. Vambery (2008) "How technological determinism shapes international marketing." *International Business & Economics Research Journal (IBER)* 7 (1): 19–28.

Garcia, Flavio D., and Bart Jacobs (2011) "Privacy-friendly energy-metering via homomorphic encryption." In *International Workshop on Security and Trust Management*, eds. Jorge Cuellar, Javier Lopez, Gilles Barthe, and Alexander Pretschner, 226–238. Berlin: Springer.

Genus, Audley, and Andy Stirling (2018) "Collingridge and the dilemma of control: Towards responsible and accountable innovation." *Research Policy* 47 (1): 61–69.

Glod, William (2015) "How nudges often fail to treat people according to their own preferences." *Social Theory and Practice* 41 (4): 599–617.

Goldin, Jacob, and Daniel Reck (2020) "Revealed-preference analysis with framing effects." *Journal of Political Economy* 128 (7): 2759–2795.

Greenberg, Andy (2017) "Good news: Android's huge security problem is getting less huge." *WIRED*, March 22. https://www.wired.com/2017/03/good-news-androids-huge-security-problem-getting-less-huge/

Greer, Evan (2022) "America's Ring doorbell camera obsession highlights the scourge of mass surveillance." *NBC News*, November 2. https://www.nbcnews.com/think/opinion/amazons-ring-doorbell-videos-make-america-less-safe-crime-rcna55143

Guffroy, Marine, Vigouroux Nadine, Christophe Kolski, Frédéric Vella, and Philippe Teutsch (2017) "From human-centered design to disabled user and ecosystem-centered design in case of assistive interactive systems." *International Journal of Sociotechnology and Knowledge Development (IJSKD)* 9 (4): 28–42.

Harris Jr., Charles E., Michael S. Pritchard, Ray W. James, Elaine E. Englehardt, and Michael J. Rabins (2013) *Engineering Ethics: Concepts and Cases*. Boston, MA: Cengage Learning.

Harrison, Natalie (n.d.) "Technological determinism: A critique based on several readings in adult education." *Learning Tech*. https://sites.psu.edu/natalieharp/writings/technological-determinism-a-critique-based-on-several-readings-in-adult-education/

Heilbroner, Robert L. (1967) "Do machines make history?" *Technology and Culture* 8: 335–345.

Henigan, Dennis A. (2016) *"Guns Don't Kill People, People Kill People": And Other Myths About Guns and Gun Control*. Boston, MA: Beacon Press.

Henry, Nicola, and Anastasia Powell (2016) "Sexual violence in the digital age: The scope and limits of criminal law." *Social & Legal Studies* 25 (4): 397–418.

Hirst, Martin (2012) "One tweet does not a revolution make: Technological determinism, media and social change." *Global Media Journal – Australian Edition* 6 (2): 1–11.

Holeman, Isaac, and Dianna Kane (2020) "Human-centered design for global health equity." *Information Technology for Development* 26 (3): 477–505.

Homer (2019) *The Odyssey*, trans. I. Johnston. Peterborough, Canada: Broadview Press.

Hotten, Russell (2015) "Volkswagen: The scandal explained." *BBC News*, 10 December. https://www.bbc.co.uk/news/business-34324772

Huesemann, Michael, and Joyce Huesemann (2011) *Techno-Fix: Why Technology Won't Save Us or the Environment*. Gabriola Island, BC: New Society Publishers.

Huxley, Aldous (2013) *Brave New World*. Berlin: Cornelian Schulverlage.

Ihde, Don, and Lambros Malafouris (2018) "*Homo Faber* revisited: Postphenomenology and material engagement theory." *Philosophy & Technology* 32 (2): 195–214.

Iqbal, Manor (2022) "Zoom revenue and usage statistics."

Business of Apps. https://www.businessofapps.com/data/zoom-statistics/

Jones, Peter (2018) "Contexts of co-creation: Designing with system stakeholders." In *Systemic Design*, eds. Peter Jones and Kyoichi Kijima, 3–52. Tokyo: Springer.

Karp, Aaron (2018) "Estimating global civilian-held firearms numbers." *Small Arms Survey*.

Kline, Ronald R. (2001) "Technological determinism." *International Encyclopedia of the Social & Behavioral Sciences*: 15495–15498.

Knight, Will, and Paresh Dave (2023) "In sudden alarm, tech doyens call for a pause on ChatGPT." *Wired*. https://www.wired.com/story/chatgpt-pause-ai-experiments-open-letter/

Kranzberg, Melvin (1986) "Technology and history: 'Kranzberg's Laws.'" *Technology and Culture* 27 (3): 544–560.

Latour, Bruno (1999) *Pandora's Hope*. Cambridge, MA: Harvard University Press.

Lavi, Michal (2018) "Evil nudges." *Vanderbilt Journal of Entertainment & Technology* 21: 1.

Logler, Nick, Daisy Yoo, and Batya Friedman (2018) "Metaphor cards: A how-to-guide for making and using a generative metaphorical design toolkit." *Proceedings of the 2018 Designing Interactive Systems Conference*: 1373–1386.

Lonergan, Bernard (1992) *Insight: A Study of Human Understanding, vol. 3, Collected Works of Bernard Lonergan*, Toronto: University of Toronto Press.

López-Meneses, Eloy, Esteban Vázquez-Cano, Mariana-Daniela González-Zamar, and Emilio Abad-Segura (2020) "Socioeconomic effects in cyberbullying: Global research trends in the educational context." *International Journal of Environmental Research and Public Health* 17 (12): 4369.

Mace, Ronald L. (1990) *Accessible Environments:*

Toward Universal Design. Raleigh, NC: Center for Universal Design.

Manancourt, Vincent (2020) "Norway suspends contact-tracing app over privacy concerns." *Politico*, July 15. https://www.politico.eu/article/norway-suspends-contact-tracing-app-over-privacy-concerns/

Maréchal, Nathalie (2017) "Global inequality in your pocket: How cheap smartphones and lax policies leave us vulnerable to hacking." *Advox*, March 30. https://advox.globalvoices.org/2017/03/30/global-inequality-in-your-pocket-how-cheap-smartphones-and-lax-policies-leave-us-vulnerable-to-hacking/

Marx, Leo (1994) "The idea of 'technology' and postmodern pessimism." In *Does Technology Drive History? The Dilemma of Technological Determinism*, eds. Yaron Ezrahi, Everett Mendelsohn, and Howard Segal, 11–28. Dordrecht: Springer.

McDaniel, Justine (2022) "They got a Ring doorbell alert, then opened fire on a bystander, police say." *Washington Post*, October 19. https://www.washingtonpost.com/nation/2022/10/19/ring-doorbell-camera-shooting-florida/

McKinney, Dave (2018) "Land of Lincoln governor goofs Honest Abe quote." Illinois Public Media, February 2. https://will.illinois.edu/news/story/land-of-lincoln-governor-goofs-honest-abe-quote

McLuhan, Marshall (1994) *Understanding Media: The Extensions of Man*. Berkeley, CA: Gingko Press.

McMillan, Caroline (2019) "Envisioning value-rich design for IoT wearables." In Loughborough University. *Proceedings of the 2nd International Conference Textile Intersections*: 1–8.

Mees, Carolin (2018) *Participatory Design and Self-Building in Shared Urban Open Spaces: Community Gardens and Casitas in New York City*. Springer.

Morton, Daniel (2015) *The Planet Remade: How*

Geoengineering Could Change the World. Princeton, NJ: Princeton University Press.

Namara, Moses, Daricia Wilkinson, Kelly Caine, and Bart P. Knijnenburg (2020) "Emotional and practical considerations towards the adoption and abandonment of vpns as a privacy-enhancing technology." *Proceedings on Privacy Enhancing Technologies* 1: 83–102.

Nathan, Lisa P., Batya Friedman, Predrag Klasnja, Shaun K. Kane, and Jessica K. Miller (2008) "Envisioning systemic effects on persons and society throughout interactive system design." *Proceedings of the 7th ACM Conference on Designing Interactive Systems*: 1–10.

Nimmanterdwong, Zethapong, Suchaya Boonviriya, and Pisit Tangkijvanich (2022) "Human-centered design of mobile health apps for older adults: Systematic review and narrative synthesis." *JMIR mHealth and uHealth* 10 (1): e29512.

Norman, Don (1988) *The Design of Everyday Things*. New York: Basic Books.

Orlikowski, Wanda J. (2000) "Using technology and constituting structures: A practice lens for studying technology in organizations." *Organization Science* 11 (4): 404–428.

Pain, Elisabeth (2017) "To be a responsible researcher, reach out and listen." *Science*. https://www.science.org/content/article/be-responsible-researcher-reach-out-and-listen

Patrick, Vanessa M., and Candice R. Hollenbeck (2021) "Designing for all: Consumer response to inclusive design." *Journal of Consumer Psychology* 31 (2): 360–381.

Persson, Hans, Henrik Åhman, Alexander Arvei Yngling, and Jan Gulliksen (2015) "Universal design, inclusive design, accessible design, design for all: different concepts – one goal? On the concept of accessibility – historical, methodological and philosophical aspects." *Universal Access in the Information Society* 14 (4): 505–526.

Peters, Adele (2015) ""Vote With Your Butt" is a brilliant idea to stop litter." *Fast Economy*, August 12. https://www.fastcompany.com/3053880/vote-with-your-butt-is-a-brilliant-idea-to-stop-litter

Peterson, Martin, and Andreas Spahn (2011) "Can technological artefacts be moral agents?" *Science and Engineering Ethics* 17 (3): 411–424.

Pinch, Trevor J., and Wiebe E. Bijker (1984) "The social construction of facts and artefacts: Or how the sociology of science and the sociology of technology might benefit each other." *Social Studies of Science* 14 (3): 399–441.

Postman, Neil (1995) "Neil Postman on cyberspace." http://www.youtube.com/watch?v=49rcVQ1vFAY&feature=youtube_gdata_player

Prainsack, Barbara (2020) "The value of healthcare data: To nudge, or not?" *Policy Studies* 41 (5): 547–562.

Raggio, Olga (1958) "The myth of Prometheus: Its survival and metamorphoses up to the eighteenth century." *Journal of the Warburg and Courtauld Institute*s 21 (1/2): 44–62.

Ranftler, David (2016) "Design thinking history – the impact of Stanford Prof. John Arnold." *Design Thinking Research*, January 30. https://blog.rwth-aachen.de/designthinking/2016/01/30/design-thinking-history-the-impact-of-stanford-prof-john-arnold/

Ross, Casey, and Ike Swetlitz (2018) "IBM's Watson supercomputer recommended 'unsafe and incorrect' cancer treatments, internal documents show." *STAT*. https://www.statnews.com/2018/07/25/ibm-watson-recommended-unsafe-incorrect-treatments/

Ruth, Matthias, and Stefan Goessling-Reisemann (2019) *Handbook on Resilience of Socio-Technical Systems*. Cheltenham: Edward Elgar.

Sari, Arif, Zakria Abdul Qayyum, and Onder Onursal (2017) "The dark side of the China: The government, society and the Great Cannon." *Society for Science and Education United Kingdom* 5 (6): 48–61.

Scheffler, Samuel (1995) "Individual responsibility in a global age." *Social Philosophy and Policy* 12 (1): 219–236.

Shahriari, Kyarash, and Mana Shahriari (2017) "IEEE standard review – Ethically aligned design: A vision for prioritizing human wellbeing with artificial intelligence and autonomous systems." In *IEEE Canada International Humanitarian Technology Conference (IHTC)*: 197–201.

Smallman, Melanie (2018) "Citizen science and responsible research and innovation." In *Citizen Science: Innovation in Open Science, Society and Policy*, eds. Susanne Hecker, Muki Haklay, Anne Bowser, Zen Makuch, Johannes Vogel, and Aletta Bonn, 241–253. London: UCL Press.

Snyder, Timothy (2018) *The Road to Unfreedom: Russia, Europe, America.* New York: Tim Duggan Books.

Sorgner, Stefan Lorenz (2021) *We Have Always Been Cyborgs: Digital Data, Gene Technologies, and an Ethics of Transhumanism.* Bristol, UK: Bristol University Press.

Stamouli, Nektaria (2022) "Hospitals refused to treat toddler because his parents were unvaccinated." *Politico*, January 31. https://www.politico.eu/article/hospitals-refused-to-treat-toddler-as-his-parents-were-unvaccinated/

Steen, Marc (2012) "Human-centered design as a fragile encounter." *Design Issues* 28 (1): 72–80.

Sunstein, Cass R. (2017) *Human Agency and Behavioral Economics: Nudging Fast and Slow.* London: Palgrave Macmillan.

Sunstein, Cass R., and Richard H. Thaler (2008) *Nudge:*

Improving Decisions about Health, Wealth, and Happiness. New Haven, CT: Yale University Press.

Tenner, Edward (2006) "Just say no to tech determinism." *Issues in Science and Technology* 23 (1): 88–93.

Thaler, Richard H., Cass R. Sunstein, and John P. Balz (2012) "Choice architecture." In *The Behavioral Foundations of Public Policy,* eds. Eldar Shafir, 428–439. Princeton, NJ: Princeton University Press.

Thinyane, Mamello, Karthik Bhat, Lauri Goldkind, and Vikram Kamath Cannanure (2018) "Critical participatory design: reflections on engagement and empowerment in a case of a community-based organization." *Proceedings of the 15th Participatory Design Conference: Full Papers – Volume 1*: 1–10.

Tolkien, J. R. R. (2004) *The Fellowship of the Ring*. London: HarperCollins.

Tolkien, J. R. R. (2007)*The Silmarillion*. London: HarperCollins.

Tollon, Fabio (2022). "Artifacts and affordances: From designed properties to possibilities for action." *AI & Society* 37: 239–248.

UK Parliament (n.d.) "Churchill and the Commons Chamber." *UK Parliament*. https://www.parliament.uk/about/living-heritage/building/palace/architecture/palacestructure/churchill/

Umbrello, Steven (2021) "Conceptualizing policy in value sensitive design: A machine ethics approach." In *Machine Law, Ethics, and Morality in the Age of Artificial Intelligence*, ed. Steven Thompson, 108–125. IGI Global, Hershey.

Umbrello, Steven (2022a) "Designing genetic engineering technologies for human values." *Etica & Politica / Ethics & Politics* 24 (2): 481–510.

Umbrello, Steven (2022b) *Designed for Death: Controlling Killer Robots*. Budapest: Trivent Publishing.

Umbrello, Steven (2022c) "The role of engineers in harmonising human values for AI systems design." *Journal of Responsible Technology* 10 (2022d): 100031.

Umbrello, Steven (2023a). *Oggetti Buoni. Come costruire una tecnologia sensibile ai valori*, Rome: Fandango.

Umbrello, Steven (2023b) "Sociotechnical infrastructures of dominion in Stefan L. Sorgner's *We Have Always Been Cyborgs*." *Etica & Politica / Ethics & Politics.*

Umbrello, Steven (2023c) "From subjectivity to objectivity: Bernard Lonergan's philosophy as a grounding for value sensitive design." *Scienza & Filosofia* 29: 36–46.

Umbrello, Steven, and Ibo van de Poel (2021) "Mapping value sensitive design onto AI for social good principles." *AI and Ethics* 1 (3): 283–296.

Van de Poel, Ibo (2013). "Translating values into design requirements." In *Philosophy and Engineering: Reflections on Practice, Principles and Process*, eds. Diane P. Michelfelder, Natasha McCarthy, and David E. Goldberg, 253–266. Springer, Dordrecht.

Van den Hoven, Jeroen (2007) "ICT and value sensitive design." In *The Information Society: Innovation, Legitimacy, Ethics and Democracy*, eds. Philippe Goujon et al., 67–73. Dordrecht: Springer.

Van den Hoven, Jeroen (2014) "Responsible innovation: A new look at technology and ethics." In *Responsible Innovation 1: Innovative Solutions for Global Issues*, eds. Jeroen van den Hoven, Neelke Doorn, Tsjalling Swierstra, Bert-Jaap Koops, and Henny Romijn, 3–13. Dordrecht: Springer.

Van den Hoven, Jeroen (2017) "The design turn in applied ethics." In *Designing in Ethics*, eds. Jeroen van den Hoven, Seumas Miller, and Thomas Pogge, 11–31. Cambridge: Cambridge University Press.

Van den Hoven, Jeroen, Gert-Jan Lokhorst, and Ibo van de Poel (2012) "Engineering and the problem of moral overload." *Science and Engineering Ethics* 18 (1): 143–155.

Van der Velden, Maja, and Christina Mörtberg (2015) "Participatory design and design for values." In *Handbook of Ethics, Values, and Technological Design: Sources, Theory, Values and Application Domains*, eds. Jeroen van den Hoven, Pieter E. Vermaas, and Ibo van de Poel, 41–66. Dordrecht: Springer.

Verbeek, Peter-Paul (2005) *What Things Do*. University Park, PA: Pennsylvania State University Press.

Von Schomberg, René (2013). "A vision of responsible research and innovation." *Responsible Innovation. Responsible Innovation: Managing the Responsible Emergence of Science and Innovation in Society*, eds. Richard Owen, John Bessant, and Maggy Heintz, 51–74. Hoboken, NJ: John Wiley & Sons.

Wakefield, Jonny (2022) "Unvaccinated woman facing removal from transplant list appeals court ruling backing hospital." *Edmonton Journal*, August 16. https://edmontonjournal.com/news/crime/unvaccinated-woman-facing-removal-from-transplant-list-appeals-court-ruling-backing-hospital

Watkins, Kari Edison, Brian Ferris, Yegor Malinovskiy, and Alan Borning (2013) "Beyond context-sensitive solutions: Using value-sensitive design to identify needed transit information tools. *Third International Conference on Urban Public Transportation Systems*: 296–308.

White, Lynn (1962) *Medieval Technology and Social Change*. New York: Oxford University Press.

Winner, Langdon (1977) *Autonomous Technology: Technics-out-of-control as a Theme in Political Thought*. Cambridge, MA: MIT Press.

Winner, Langdon (1980) "Do artifacts have politics?" *Daedalus* 109 (1): 121–136.

Winner, Langdon (1993) "Upon opening the black box and finding it empty: Social constructivism and the philosophy of technology." *Science, Technology, & Human Values* 18 (3): 362–378.

Winner, Langdon (2020) *The Whale and the Reactor A Search for Limits in an Age of High Technology*, 2nd edn. Chicago: University of Chicago Press.

Wolff, Jonathan (2020) "Fighting risk with risk: Solar radiation management, regulatory drift, and minimal justice." *Critical Review of International Social and Political Philosophy* 23 (5): 564–583.

Wyatt, Sally (2008) "Technological determinism is dead; Long live technological determinism." In *The Handbook of Science and Technology Studies*. 3rd edn., eds. Edward J. Hackett et al., 165–180. Cambridge, MA: MIT Press.

Yoo, Daisy (2021) "Stakeholder tokens: A constructive method for value sensitive design stakeholder analysis." *Ethics and Information Technology* 23 (1): 63–67.

Yoo, Daisy, Katie Derthick, Shaghayegh Ghassemian, Jean Hakizimana, Brian Gill, and Batya Friedman (2016) "Multi-lifespan design thinking: Two methods and a case study with the Rwandan diaspora." *Proceedings of the 2016 CHI Conference on Human Factors in Computing Systems*: 4423–4434.

Yoo, Daisy, Nick Logler, Stephanie Ballard, and Batya Friedman (2022) "Multi-lifespan envisioning cards: Journeying from design theory to tools for action." In *Designing Interactive Systems Conference*, 557–570.

Index